The Economics of European Agriculture

The Economics of European Agriculture

Bernadette Andreosso-O'Callaghan

Professor of Economics
University of Limerick
Ireland

palgrave
macmillan

First published 2003 by
PALGRAVE MACMILLAN
Houndmills, Basingstoke, Hampshire RG21 6XS and
175 Fifth Avenue, New York, N.Y. 10010
Companies and representatives throughout the world

PALGRAVE MACMILLAN is the global academic imprint of the Palgrave Macmillan division of St. Martin's Press, LLC and of Palgrave Macmillan Ltd. Macmillan® is a registered trademark in the United States, United Kingdom and other countries. Palgrave is a registered trademark in the European Union and other countries.

ISBN 0–333–72644–8

This book is printed on paper suitable for recycling and made from fully managed and sustained forest sources.

A catalogue record for this book is available from the British Library.

Library of Congress Cataloging-in-Publication Data
Andreosso-O'Callaghan, Bernadette, 1959–
The economics of European agriculture / Bernadette Andreosso-O'Callaghan.
p. cm.
Includes bibliographical references and index.
ISBN 0–333–72644–8
1. Agriculture—Economic aspects—Europe. I. Title.

HD1917.A63 2003
338.1'094—dc21

2003048287

10 9 8 7 6 5 4 3 2 1
12 11 10 09 08 07 06 05 04 03

Printed and bound in Great Britain by
Antony Rowe Ltd, Chippenham and Eastbourne

A Maman

Contents

List of Figures

List of Tables

Acknowledgements

During the gestation and the preparation of this book, there were a number of people who, in one way or another, contributed to it by stimulating me to delve into the intricacies of European agricultural economics. First of all, my teaching to several cohorts of students at the University of Limerick, and at the University of Montpellier (France) where I have been invited Professor, gave me the opportunity to lay down the foundations of this work. Had it not been for these students, I may never have written such a book. My thanks go to my research students, and in particular to Patrick Callinan for his invaluable help, and I am grateful to Catherine Sexton for the efficient support she provided in relation to the graphs. I am extremely indebted to Michael Tracy who taught Agricultural Economics to me at the College of Europe and to Jean-Pascal Bassino, Sylvie Vacheret and Rita Buckley for their comments and constructive criticisms proffered on an earlier version of the book. Invaluable assistance was provided by Marie Dineen from the European Documentation Centre at the University of Limerick at all stages of the writing process.

Last but not least, I would like to thank my husband Frank O'Callaghan and our three children for their infinite patience, support and love. In particular, Frank has kindly proofread large sections of the manuscript.

Permission to reproduce copyright material is greatly acknowledged in particular to the Econometrics Society. Also, I am grateful to Cambridge University Press for allowing me to reprint the following: Tables 1.2, 5.2 and 12.1 of James Simpson's book entitled *Spanish Agriculture – The Long Siesta, 1765–1965*, published in 1995.

Introduction

In an era molded by the dominance of the so-called 'new economy', that is, the growth of the information and computer technology sector, and also by the gradual and inexorable decline of the agricultural sector as a share of total human economic activities, one may wonder what is the rationale for another book on agricultural economics. Besides the usual arguments, which can be summarized by the idea that agricultural economics deals with questions vital for the well-being of human kind, for example, the production of food, a basic necessity, and that the capabilities of agricultural economists can best be used for resolving real-world problems, two important dimensions have emerged in recent years. First, the recent advances in food engineering, through the production of genetically modified organisms (GMOs), and their perceived risks on human beings have revived the debate on food quality, and on consumer safety and welfare. The interest in food quality has also been fuelled by recent outbreaks of bovine spongiform encephalopathy (BSE) and foot and mouth diseases in Europe. Second, food security, that is, the ability of agriculture to provide adequate food supplies, coupled with a growing population and the intensification of agricultural productive methods, which through pollution and environmental degradation undermine future agricultural productivity, are important and growing problems facing humankind in this new millennium. A major concern for many countries in this new century will indeed be the ability to find sufficient and adequate water supplies for the production of food commodities.

This book is centered on the case of the European Union (EU).[1] One of the major reasons for this is that over the course of its short history, the EU has established itself as a major world producer and trader of agricultural commodities. Numerous studies have shown how the

Common Agricultural Policy (CAP) has created tensions at the world level, how it has been perceived as being inequitable both within the EU and on world markets, and how it has contributed to reducing world prices and to raising entry barriers to producers from least developed economies of the world. To these numerous criticisms, the EU has responded by a re-ordering of the CAP through the reforms of the 1990s. The last decade of the 20th century has also been characterized by a pronounced shift in attitude in terms of EU economic policy. The EU has become more involved in collaborating with third countries (in particular with third-world countries), as it aspires to become for example an important actor in food research with the developing world. This book therefore emphasizes a different angle of EU agricultural policy, from the one that can normally be found in similar texts. Through extensive coverage of the EU agricultural relations with developing countries, it attempts at delineating a new and emerging architecture for a world agricultural system where the EU would play a major role. Moreover, because of recent advances in the area of innovation in agriculture, it is now increasingly difficult to separate the field of agriculture from that of manufacturing. The frontier between the two sectors has become increasingly blurred. An example of this is that the top firms involved in food technology research, that is, Novartis, Monsanto, Du Pont de Nemours, Zeneca, AgrEvo, Rhône Poulenc, Bayer, American Cyanamid, Dow Agrosciences and BASF are not food companies, but rather large chemical complexes. Consequently, the book integrates what goes normally under agricultural economics with other areas of economics, by devoting a full chapter to the issues of innovation and new growth theories.

The structure of the book is as follows: Chapter 1 introduces the main concepts used throughout the book, such as those of 'agricultural system' and 'filière'. It also suggests an overview of the main trends in agricultural supply, and it concludes with a discussion on the role of agriculture in economic development with due reference to theories of agricultural development. Chapter 2 introduces the tools of demand and supply so as to allow the reader to understand the market mechanisms for agricultural products. The description of concepts and intellectual constructs such as 'elasticity', 'productivity' and 'cobweb theorem' leads to a discussion of the 'farm problem'. Chapter 3 is devoted to Government intervention in the field of agriculture. The approach taken is historical, as the nature of CAP can only be understood by reference to government intervention in the agricultural sectors of individual post-war European economies, and in particular, of the six founding members. The chapter goes on to highlight the major policy instruments. Chapter 4 follows by focusing on CAP, its genesis, its objectives and instruments, as well as its results. The 1992 CAP reform and Agenda 2000 conclude this chapter. A historical perspective is also used in Chapter 5, which is entitled 'Technology and Innovation'. After an insight into the association between technology and economic growth, a brief account of the major agricultural innovations in Europe is provided. Chapter 5 also examines some theoretical models of innovation diffusion, as well as the importance of knowledge in agriculture, by focusing on the notion of agricultural knowledge system (AKS). It concludes by devoting a section to agricultural innovation as an EU policy. Chapter 6 covers the topic of the enlargement, in a dynamic perspective. The fifth enlargement, or the integration of former centrally planned economies of Eastern Europe, is the main focus of the chapter. Chapter 7 looks at the EU as an agricultural system in the world. By focusing on trade, the chapter reviews the main trade arrangements signed between the EU and other developing nations of the world, such as the African Caribbean and Pacific states (ACP). Special attention is also devoted to the problem of poverty and under-nourishment in developing nations, and to the EU policies aimed at tackling these serious difficulties. Finally, trade liberalization in agriculture is the central issue of Chapter 8. The negotiating position of the EU in the Uruguay Round is presented, the CAP–GATT link is assessed, and the impact of trade re-ordering is attempted. The last section of the chapter deals with the major issues favored by the EU in the framework of the new WTO round.

Recommended reading

Hennig Hanf, C. (1997) 'Agricultural economics in Europe: a thriving science for a shrinking sector', *European Review of Agricultural Economics*, 24/3–4, 565–78.

Note

1 The European Union has evolved from being a Common Market in 1958, to becoming a European Economic and Monetary Union (EMU) established during the 1990s. The Common Market was founded by six member countries: Belgium, France, Germany, Italy, Luxembourg and the Netherlands. Another nine European countries have joined since the early 1970s, namely: Denmark, Ireland, the UK, Greece, Portugal, Spain, Austria, Finland and Sweden.

1
Concepts and Definitions

Objectives of this chapter

- To introduce the basic concepts necessary for the understanding of the sphere of agricultural economics.
- To review the major trends in agricultural supply at both EU and world levels.
- To study the role of agriculture in economic development.

1.1 Definition and scope of the field of agricultural economics

The field of agricultural economics is delineated by the application of economic science tools to the agricultural sector. Economics is the science of the administration of scarce resources (land, labor, capital and management), which are needed in order to produce goods and services that satisfy human wants. Agricultural economics refers to all economic activities connected with the control of living organisms, such as plants and animals. These economic activities gravitate around the production of food, and they involve many different economic actors at different production and transformation stages. The agricultural economist is interested in the process that leads ultimately to the satisfaction of human wants. This process encompasses the conditions of production, the characteristics and evolution of demand, the mechanisms prevailing on agricultural markets, government intervention and world trade conditions.

1.1.1 Early views on the scope and role of agriculture

Attempts to define clearly the boundaries of the agricultural sector go back to 1758 with François Quesnay's publication of the famous *Tableau*

Economique. Leader of a new doctrine emerging in France at the time and called the physiocratic[1] movement, Quesnay represented the economy as a circular flow in which the natural laws were predominant. Quesnay's economic table shows three sectors of economic activity: (i) the farming sector, which he calls the 'productive class'; (ii) the sector of landowners; and (iii) the 'sterile' class or sector comprising artisans, servants and the clergy. According to Quesnay, the farming sector is the only sector that can produce an output greater than its cost of production; therefore farming activities, representing a dominant sector at the time, were accorded a pivotal role in the physiocratic doctrine. What Quesnay and his disciples had in mind was to show scientifically, that is, with simple algebra, that the manufacturing and trading sectors were non-productive. The demonstration of this clear idea was a very tempting tool used against mercantilism, the economic doctrine during the 16th, 17th and the first part of the 18th centuries. More importantly, in showing the interrelatedness between the different economic sectors, and in trying to quantify their magnitude, Quesnay's economic table represents a major methodological advance in the field of economics. Closer in time to us, Fisher (1939) and Clark (1940) modernized the classification of economic sectors in introducing services as a new category of economic activity, alongside agriculture and manufacturing. Moreover, among the most elaborate modern versions of Quesnay's economic table is Leontief's input–output table that paved the way to quantitative economics and econometrics (Leontief, 1951). In particular, the matrix of intermediate consumptions (or inter-industry matrix) of a country's input–output table depicts the pattern of transactions existing between the different industries during a given year. Table 1.1 represents a simplified version of a hypothetical inter-industry matrix for the EU as a whole, where Agriculture/Forestry and Fishing appears as one distinct industry. Industries belonging to the manufacturing sector would be located around industry j of our table, whereas the services sector would appear in the vicinity of industry n. For purpose of simplicity, Table 1.1 does not show the matrix of final demand and other transactions such as imports, taxes, wages, depreciation and profits that would give a more comprehensive picture of the input–output table.

What it shows are the sales of domestically produced goods and services to those same industries. For example, the first row of the table indicates all the sales of Agriculture/Forestry and Fishing to the other domestic industries. Conversely, the first column represents all inputs bought by Agriculture/Forestry and Fishing to the other industries

Table 1.1 The inter-industry matrix of the EU (year *t*, in millions of €)

	Industry 1 agriculture, forestry and fishing	...	**Industry *j***	...	**Industry *n***	**Total inter-industry**
Product 1 agriculture, forestry and fishing	30	...	50	...	05	600
Product 2	20	...	40	...	10	500
...	...	...	...	...	...	...
Product *i*	40	...	70	...	40	900
...	...	...	...	...	...	...
Product *n*	20	...	40		10	
Total domestic Flows	400	...	800	...	300	8000

and sectors of the European economy. This column informs us that agriculture has bought a total of Ecus 400 million worth of inputs, of which 20 millions came from industry 2, 40 millions from industry *i*, 20 million from industry *n*, and 30 million from itself.

Although Leontief's table is useful in that it clearly shows the interdependence of the various sectors of the economy, it is unfortunately too aggregate for the purpose of the agricultural economist since it does not allow for a distinction between agriculture, fisheries and forestry.

It is clear that since François Quesnay the meaning and importance of agriculture has gone through evolving stages. If originally the meaning of agriculture was confined to the growing of crops and to the raising of livestock, today it is a much broader concept, given the variety and the greater sophistication of agricultural products, and the complex nature of businesses involved in agricultural activity. Modern agriculture has become a complex system. It embraces today all activities connected with the transformation of natural resources; it now involves farms, agribusinesses, governmental organizations, local and regional institutions, lobbying groups, as well as financial institutions.

1.1.2 A modern vision of the agricultural system

The economic activities related to the production of food can be grouped into four sub-sectors that can be identified by their nature and their function in the agricultural system: the farm sub-sector, the agribusiness sub-sector, the public sub-sector and the financial sub-sector.

1. The *farm sub-sector* includes all the farm-firms that grow crops and raise livestock, usually for sale. When compared with other production units in the economy, the agricultural holdings combine a number of idiosyncrasies.

First, they are small-sized, when compared with other production units in the manufacturing sector. In 1997, the average utilized agricultural area (UAA) was 18.4 hectare in the EU.[2] Obviously, this average figure masks the important structural imbalances existing in the European agricultural sector. In the same year, the average UAA of British farm holdings was more than 69 hectare, against 4.3 hectare only in Greece. In 1999, the average total agricultural output per agricultural holding, and in current terms, amounted to €170,898 in The Netherlands. This compares with only €15,129 and €13,745 in the case of Portugal and Greece, respectively.

Being a direct implication of their small size, individual farmers have little power in influencing prices in the market. In the past, this lack of power was slightly counterbalanced by the organization of farming units into co-ops, and later into politically organized farmers' association (see Chapter 3). Born in the 19th century with Robert Owen in the UK and Charles Fourier (1822) in France, the cooperative movement has been a distinctive characteristic of agricultural development in Europe and in other countries of the world. In 1997, more than 95 per cent of all milk produced in Ireland, 85 per cent of pig meat produced in France, 76 per cent of all Dutch fruits, and at least 70 per cent of all Danish fruits, vegetables, pig meat and milk, were sold through cooperatives. In Italy, Greece, Spain and Britain, the cooperatives have a more marginal role. At best, they absorb and control 43 per cent of all fruits produced in Italy, 45 per cent of fruits in Spain and 20 per cent of milk in Greece.

Second, since the crops and livestock produced by the farms are living organisms, agricultural production is subject to a high level of uncertainty reflected in quasi-unpredictable climatic conditions, disease and product life cycles. In particular, the weather still dictates distinct patterns of seasonality. Also, crops and livestock depend upon the availability of land and water. This introduces geographical and topographical limits on agricultural production. For example, the mountain regions of the EU (in Austria, Italy and France) are not suitable for intensive farming and place a limit to the variety of food that can be produced.

Third, part-time working is an important feature of employment in the farm sub-sector. According to official EU estimates (CEC, 2002) the share of farmers who were at the same time farm heads and who were working full time was only 27 per cent for the EU as a whole in 1997. Part-time farming affects more than 75 per cent of the farming population in Greece, Portugal and Spain. It is also a relatively common phenomenon in the more agriculturally advanced EU economies such as Germany (61.2 per cent), Denmark (50.8 per cent) and The Netherlands (32.7 per cent).

2. *The agribusiness sub-sector*. Following the classification suggested by Seitz *et al.* (1994), the agribusiness sub-sector can be divided into:

(i) The input sub-sector, which encompasses all the firms and industries that produce and sell goods and services that are used as inputs by the farms. These goods and services include fertilizers and pesticides (purchased from the chemical industry) machinery (purchased from the mechanical engineering industry) and animal foodstuff (from the food, drink industry). These industries are all listed under column 1 of the above inter-industry matrix.

(ii) The processing and marketing sub-sector, regroups all the firms and industries that purchase, store, process and distribute farm products either on the domestic or on export markets. This sub-sector is vast since it involves most of the firms included in the food-drink-tobacco industry that process, package and market farm commodities, the large – mostly international – trading companies, as well as part of the transport and distribution sector (supermarkets, distribution chains).

3. The *Public sub-sector* is heterogeneous and includes:

(i) Governmental agencies that provide services such as sanitary regulations, food inspection and market information and supervision.

(ii) The educational and research group represented by schools, universities and other state financed research centers which provide education, training and research exclusively to the agricultural sector. One such well-known educational institution is the Waegeningen University in The Netherlands, European center of agricultural training and research. Adult education is referred to as 'extension education'. Extension services support farmers through the dissemination of knowledge and research results, in a wide variety of subjects.

(iii) The government whose functions in this particular area often encompass price support policies, aid for foreign trade, special credit concessions and grants of all kinds. At the EU level, the Council of Ministers for Agriculture and the different working groups within the community institutions exert an important influence on the determination of agricultural prices.

(iv) The farm lobby consisting of national farmers' associations has a decisive impact in terms of national agricultural policy. Throughout Europe, the farm organizations saw their strength and influence increase after the Second World War. These associations are for example: the Deutscher Bauernverband (DBV) in Germany, the Fédération Nationale des Syndicats des Exploitants Agricoles (FNSEA) in France, the Landbouwschap in The Netherlands, the Confederazione Nazionale dei Coltivatori Diretti in Italy and the Irish Farmers Association (IFA) in Ireland. Since 1958, the farm organizations of the EU members have been united under the Comité des Organisations Professionnelles Agricoles (COPA).

4. The *financial sub-sector* has been an important component in the development of the agricultural system in Europe. During the 19th century, part of the financial sector was exclusively connected with agricultural activities. Before they diversified their clientele and products, the Crédit Agricole in France, the numerous Casse Rurale in Italy, the Agricultural Credit Corporation in Ireland, all served as the major providers of credit and soft loans to the farming community.

In order to capture the dynamics of the agricultural system in particular – and of other industrial systems in general – as well as the inter-dependent relationships between the various components of the system, the French economists have suggested an analysis in terms of *filière*.

1.1.3 The agri-food *filière*

The origin of the concept of *filière* is found in the works of Aujac (1960), de Bernis (1966) and also of François Perroux (1973). The *filière* emphasizes the technological links among related manufacturing activities encompassed in a given industrial system. At the core of this notion lies the concept of 'determinant' or 'key' industry and its transmission effects throughout the industrial system. A *filière* is thus constituted by a chain of economic activities, ranging from the extraction of natural resources to the distribution of the goods. The units of production belong to different industries linked together by buyer–supplier relationships. More broadly, the *filière* can be defined as a linking filament among technological, product and capital-based activities (Toledano, 1978). Montfort (1983) sees the *filière* as comprising three levels: upstream, center and downstream. The upstream pole is a set of industries supplying to the core industries of the *filière*. Conversely, the downstream pole consists of the industries that buy from the core industries of the *filière*, without any reciprocal relationship. By deduction, the core of the *filière* comprises all the industries involved in the transformation of inputs such as raw materials and primary products into finished goods.

The agri-food *filière* will thus comprise the following poles:

- The upstream pole includes farming and fisheries activities, agricultural machinery and animal feed producers.
- The downstream pole comprises catering services, food haulage and distribution of finished products.
- The core consists thus of the food processing industry itself.

Developed in the 1970s in the context of declining competitiveness of the French industry, the notion of *filière* is suggestive of nationalistic and interventionist ideas. In the case of the French economy, 19 *filières*, among which was the agri-food *filière*, were defined on the basis of the relationships appraised in the input–output tables (Angelier, 1991). The different segments of the *filière* were examined, for example in terms of import penetration ratios. A high import penetration ratio in the upstream or core poles would have repercussions for the whole *filière* and would diminish national industrial independence. The implications in terms of economic policy were clear: to select and sustain important *filières* within which to build up a strong industrial complex of competitive firms capable not only to recapture the domestic market, but also to gain increasing market shares on the world market.

1.2 Global trends in supply

1.2.1 World food supply and population[3]

World food production has exceeded population growth since the 1970s. In particular, over the 1990–99 decade, the population of the world has increased by 13.5 per cent, but world food production has grown even faster, by 20 per cent. The breakdown by major regions in the world shows that this pattern has been normal and universal, although the African continent was an exception to this trend during the 1980s. Indeed, during the 1980s, food production in Africa increased roughly by 32 per cent, whereas the population of that continent had increased by 35 per cent. In particular, sub-Saharan Africa is a region in the world where per capita food production has declined in the last few decades. In the early 1990s, the growth rate of world food supply was declining slightly. The breakdown by major world regions shows that only the economically advanced regions of the world (USA, Western Europe, Australia/New Zealand), as well as the former USSR, have experienced a downward trend in food production. The decline was particularly noticeable in the USA since the mid-1980s. The contraction of food production in the developed world is a direct result of the reordering of world agricultural markets, reordering that was initiated by the last GATT negotiations (see Chapter 8). In short, food production is being curbed in order to minimize the costs of market intervention. In particular, the reform of the Common Agricultural Policy (CAP) is deemed necessary to facilitate the disposal of large surpluses in the EU. In contrast, food production in the less developed regions of the world has continued on its ascendant path, especially in Asia where it increased by more than 40 per cent between 1990 and 1999. Overall, the developing world is gradually catching up with its richer counterparts, where food supply has developed into costly surpluses. The availability of better technologies (irrigation techniques, machinery, fertilizers), combined with the dissemination of extension services, explains the uninterrupted food production growth that has taken place in the developing countries. As a result, the world increasingly produces enough food for the entire growing population of our planet. This observation goes against the thesis of Robert Malthus (1803), who argued that because populations are sets of biological organisms, who increase *ceteris paribus*[4] in geometric progression, and because resources in general, and in particular supplies of food, tend to grow in arithmetic progression, a slower population growth would become inevitable. In the British scholar's view, agriculture would be unable to keep pace with population growth trends. Clearly, the spectacular impact of technological change on agricultural productivity was totally unheard off, unexpected and unseen in early 19th century Britain. At best, the moral philosophers of the then powerful British Empire, faced with the industrial revolution, were only starting to argue on the merits and disadvantages of mechanization in the manufacturing sector.

Table 1.2 depicts the improvement in economic welfare as defined by the number of calories received by each individual since the 1960s. After the Second World War, the improvement in peoples' nutrition

Table 1.2 Food supply – calories per caput per day

	1961–63	**1988–90**
World	2287	2697
Africa	2155	2348
North Central America	2858	3333
USA	3067	3642
South America	2391	2624
Asia	1888	2494
Europe	3088	3452
Australia and New Zealand	2975	3173

Source: FAO (1991) Yearbook Production, Vol. 45.

standards has been substantial. Much of this success is explained by the *green revolution* that took place in the 1960s. The introduction of chemical fertilizers and pesticides, as well as remarkable breakthroughs in biotechnology, with the creation of high-yielding hybrid varieties of wheat and maize made this *revolution* possible. For example, India was able to double its average yield of wheat within a few years after the introduction of these developments in the late 1960s (World Development Report, 1994). According to a recent FAO report (FAO, 2002) world agriculture produces today 17 per cent more calories per person than it did 30 years ago, despite a 70 per cent population increase. It is indisputable that if the world's food production were evenly divided among the world's entire population, each individual would receive appreciably more than the minimum nutritional amount required for survival, that is at least 2720 kilocalories (kcal) per day. Table 1.2 describes a hypothetical situation: one in which the distribution of food is perfectly equal, as determined by simple arithmetic. However, in the real world, and in spite of the remarkable progress in food production and productivity achieved since the Second World War, pockets of extreme poverty and persistent famine are too often found in the poorest areas of the less developed countries, primarily in Africa, Asia and Latin America. It is estimated that a substantial part of the world population (that is, more than 800 million people) still suffers from severe malnutrition today. More worryingly, although malnutrition and famines have generally receded in recent decades, new evidence shows that, when China is excluded from the statistics, the number of undernourished people has actually increased during the 1990s. Clearly, hunger and poverty are related to the unequal distribution of incomes and wealth within countries, rather than to the scarcity of food supply at the global level, a problem which has not been given enough attention at least until recent years (see Chapter 8).

1.2.2 Major world food producers and products

As shown in Table 1.3, Asian countries are by far the major producers of food in the world in volume terms. With nearly half of world production, this group of countries outpaces by far North America, and in particular the USA, the biggest single producer and exporter of food in the developed world. The world share of Europe (including the EFTA countries) in terms of food production is approximately 15 per cent. This share is more or less equivalent to that of North America. Special attention should also be paid to the former USSR, who with nearly 10 per cent of world production in 1991 was at the time the fourth major producer (FAO, 1991).

A large part of world food production involves only a few product categories. Indeed, cereals and milk represent more than half of the total production in volume terms, and as we will see in Chapter 7, only a small proportion of food produced each year is actually traded on the international market. The share of the broad regions in world production of agricultural commodities is depicted in Table 1.4.

This table gives an indication of the *revealed comparative advantages* of the various regions of the world in terms of agricultural production. Some regions appear as the world specialist areas for the production

Table 1.3 World food production in volume (1999)

Region	(1000 Mt)	% of world total
North and Central America	833,936	16.1
Europe*	777,286	15.0
Oceania~	84,641	1.6
Africa	448,478	8.6
Asia	2,333,719	45.0
South America	404,072	7.8
Others#	304,473	5.9
World	5,186,605	100.0

* This includes EU and EFTA countries (Norway, Iceland, Switzerland, Liechtenstein).
~ That is, Australia and New Zealand.
# This group encompasses estimates for the Russian Federation as well as for the independent republics of the former USSR such as Belarus, Estonia, Latvia, Lithuania, Moldova and Ukraine. Are also included here Middle East countries.
Source: FAO (2001).

Table 1.4 Major world producers in 1999 – breakdown by product category (in %)

	North and Central America	Europe	Oceania	Africa	Asia	South America	Others
Total cereals	20.5	14.0	1.6	5.5	48.3	4.8	5.3
Root crops	4.7	12.0	0.6	24.1	42.0	7.1	9.5
Pulses	12.0	10.7	3.9	13.3	50.2	6.5	3.4
Vegetables	7.9	11.7	0.5	6.6	66.2	2.7	4.4
Fruits	11.9	15.8	1.3	13.4	39.9	15.9	1.8
Nuts	19.1	14.9	0.6	11.7	47.3	3.7	2.7
Oil crops	29.6	8.0	1.3	4.8	33.5	19.6	3.2
Sugar	15.3	16.5	4.7	6.7	32.7	21.1	3.0
Cocoa beans	2.9	–	1.6	66.1	16.1	13.3	0.0
Coffee	17.5	–	1.0	19.0	22.6	39.7	0.0
Tea	0.0	–	0.2	12.6	82.4	2.2	2.6
Vegetable fibers	20.2	2.8	3.0	7.5	53.5	5.3	7.7
Tobacco	11.0	7.4	0.1	6.6	61.8	11.5	1.6
Natural rubber	0.8	–	0.1	5.0	93.1	0.1	0.9
Total meat	20.8	20.0	2.2	4.7	38.3	10.1	3.9
Total milk	16.8	28.5	3.8	4.7	26.3	8.2	11.7
Eggs	14.9	14.2	0.5	4.4	54.6	5.6	5.8
Wool	1.2	9.8	39.4	8.9	26.2	8.3	6.2
Total	16.1	15.0	1.6	8.6	45.0	7.8	5.9

Source: FAO (2001). Author's volume based calculations.

of certain agricultural commodities. Likewise, the production of some agricultural goods is heavily concentrated on a few regions only. For example, Oceania represents only 1.6 per cent of the world total, and yet it produces more than 39 per cent of the wool of the world. In the same way, the far-east Asian countries account for more than 93 per cent of total natural rubber production and for 89 per cent of all rice produced. Cocoa beans, coffee and tea are other commodities produced by a few regions only. Latin America produces two-thirds of all coffee beans, and Africa more than half of total cocoa beans. Because of their topographical and climatic diversity, and also due to the implementation of radically different agricultural and economic policies, Northern America, Europe and the former Soviet Union present a more diversified production structure. Northern America produces 30 per cent of all coarse grain (maize essentially), whereas the EU is fairly important in the production of milk and meat at the world level.

1.2.3 EU agriculture in a global context: the structure of the EU agricultural sector, EU and world distribution of food supply[5]

In 1999, the agricultural sector of the EU employed slightly less than 7 million people; this represented roughly 4.5 per cent of total civilian employment.[6] Around 7 million agricultural holdings, with an average size of 18.4 hectares, were spread over the 136 million hectares defined as the UAA in the EU, and they were source to a final production estimated at around €274 billion. Table 1.5 depicts the share of individual products in final agricultural production. Also, it shows the relative importance of each product to the agricultural sector of major EU producers.

Table 1.5 Share of products in final EU production, and major producers for each product (1999, in %)

	EU-15	Major EU producers*
Wheat	6.2	UK, France, Denmark
Rye	0.3	Germany, Austria
Oats	0.4	Sweden, Finland
Barley	3.0	Denmark, Finland
Maize	2.3	Austria, Greece, Italy
Rice	0.3	Italy, Portugal, Spain
Sugarbeet	1.8	Belgium, Sweden
Tobacco	0.4	Greece
Olive oil	2.1	Greece, Spain, Italy
Oilseeds	2.0	France, Germany, Austria
Fruits and vegetables	14.9	Spain, Italy, Portugal
Wine and must	6.3	Portugal, France, Luxembourg
Seeds	0.4	The Netherlands, Denmark
Textile fiber	0.0	Portugal, France
Hops	0.0	Germany
Milk	13.9	Luxembourg, Sweden, Ireland
Cattle	10.2	Ireland, Luxembourg, Belgium
Pig meat	7.4	Denmark, Belgium, Austria
Sheep and goat meat	2.3	Greece, UK, Ireland
Eggs	1.6	UK, Sweden, Italy
Poultry	3.7	UK, Portugal, France
Potatoes**	2.4	UK, The Netherlands, Belgium
Other**	14.8	
Total	100.0	
Value in Mio €	268,960.0	

* Are shown only the two or three countries with the highest shares of a given product in their final agricultural production.

** Products not subject to EU market organization. The 'other' category includes mostly agricultural services.

Source: CEC (2002), T/28 and T/29.

Four major groups of products account for nearly two-thirds of total EU agricultural production: meat (19.9 per cent of total production), milk (13.9 per cent), fruits and vegetables (14.9 per cent) and cereals (12.5 per cent). These products represent essential outlets for a number of countries. In particular, the Irish agricultural sector is extremely dependent upon two major product categories: meat (essentially beef) and milk. These two product categories represent more than 58 per cent of all Irish agricultural production. Only Luxembourg has developed a specialization as narrow as that of Ireland. The agricultural sector of Luxembourg is indeed extremely specialized by European standards. In Luxembourg (a country with less than 430,000 inhabitants), milk and meat represent more than 33.6 per cent and 25.4 per cent of total national agricultural production, respectively. However, this country has also a relatively well-developed wine sector, accounting for more than 10 per cent of national agricultural production. Denmark is another small and open economy (SOE) narrowly specialized, albeit to a lesser extent than either Ireland or Luxembourg. Meat, milk as well as cereals represent nearly 67 per cent of all Danish agricultural production.

The intensity of specialization is normally an inverse relationship of a country's size. Not surprisingly, the larger member states have a more diversified agricultural sector than the smaller ones. In particular, Italy, France and Spain are relatively more diversified than their north-European counterparts. Italy's major products are fresh fruits and vegetables, wine, milk, beef/veal and olive oil. France's major products are wine, milk, beef, wheat and fresh fruits and vegetables. Spain is relatively engaged in the production of fresh fruits and vegetables, pig meat, milk, beef and veal, and olive oil. For each of the three Mediterranean countries, no product accounts for more than 15 per cent of national agricultural production, with the exception of fruits and vegetables for both Spain and Italy. The German and UK milk sectors are relatively large (20 and 17 per cent of total domestic production respectively). Other predominant products for the UK are beef, wheat, sheep and goat meat, fresh vegetables, pig meat and poultry. Besides milk, the German agricultural sector is relatively dominated by pig meat (10 per cent of national agricultural production), cattle (nearly 10 per cent) and wheat (7.8 per cent). In these larger economies, specialization patterns are looser than in SOEs.

The various EU member states show contrasted degrees of differentiation as well as slightly different specialization patterns. The south-European member states are more relatively engaged in the production of wine, fruits and vegetables, whereas the production structure of the north-European countries is relatively dominated by milk, meat and cereals (with the exception of maize). Because of its extreme geographical diversity, France combines characteristics common to the Mediterranean as well as to the north-European countries of the EU. The relative specialization patterns of the various countries, appraised at the EU level, appear in Table 1.6.

The table shows first the ranking of the major agricultural producers in the EU. With nearly a quarter of EU agricultural production, France is the biggest producer in the EU, in absolute terms. Also important producers are Italy, Germany, Spain and to a lesser extent the UK. In relative terms, that is when allowing for corrections by the size of the country (column 2), Denmark, Greece, The Netherlands and Ireland are small open EU countries with a very important agricultural sector. Spain also belongs to this first group, albeit to a lesser extent. Even when taken into account the relative size of countries, France still appears as having a relatively large and developed agricultural sector. Clearly, Germany and the UK are not important agricultural producers in relative terms.

Table 1.6 Member states' shares in final agricultural production (total and selected products; 1999, in %, EU-15 = 100)

	Population	**Total agricultural production**	**Milk**	**Beef**	**Fresh vegetables**	**Maize**
Belgium	2.7	2.5	2.6	4.1	3.6	0.1
Denmark	1.4	2.8	3.9	1.5	0.6	0.0
Germany	21.9	15.2	21.7	14.8	5.6	9.0
Greece	2.8	4.1	2.2	0.9	6.6	7.9
Spain	10.5	12.2	5.6	8.8	22.5	9.9
France	15.7	23.0	20.2	27.8	14.1	36.9
Ireland	1.0	2.0	3.7	6.5	0.8	0.0
Italy	15.3	15.4	11.0	12.7	24.5	28.4
Luxembourg	0.1	0.0	0.2	0.2	0.0	0.0
The Netherlands	4.2	6.8	8.5	5.0	8.8	0.2
Austria	2.2	2.0	2.2	1.9	0.6	4.0
Portugal	2.6	2.3	1.6	1.5	3.8	3.4
Finland	1.4	1.3	2.7	1.0	0.7	0.0
Sweden	2.4	1.6	3.0	1.3	0.6	0.0
UK	15.8	8.8	10.6	12.1	7.1	0.0

Source: CEC (2002), T/25, T/26, T/20 and T/31.

Interestingly, the table also shows how the production of some commodities is heavily concentrated on a few countries only. Two countries dominate the maize sector: France with more than one-third of the EU total, and Italy to a much lesser extent, representing another 28.4 per cent. Although not directly visible in the table, France is also dominant in the 'quality wine' sector (nearly 100 per cent of EU production), as well as in the production of oilseeds, with more than 42 per cent of EU production. Finally it is worth mentioning Germany's dominance in the rye sector (more than 76 per cent of the total EU production).

Beef is also concentrated in a small number of EU countries, although less so than in the case of cereals in general. The four largest EU countries, that is France, Germany, Italy and the UK account for more than two-thirds of EU beef production. However, the significance of the beef industry in small countries such as Ireland and, to a lesser extent, as Belgium needs to be highlighted. In spite of its small size, Ireland represents 6.5 per cent of total EU beef production. Finally, the production of vegetables is less and less concentrated in the Mediterranean countries, where natural factor endowments, that is, plentiful sunshine, explain traditional specialization patterns in this sector.

If national specialization patterns are discernible, regional differentiations ought to be taken into account, essentially in the larger member states. Table 1.7 depicts the major agricultural regions of a selected number of EU countries. These regions are those for which:

$$\left(\frac{AGVA}{TGVA}\right)_r^i \geq \left(\frac{AGVA}{TGVA}\right)_n^i$$

where:

$\left(\frac{AGVA}{TGVA}\right)_r^i$ corresponds to the share of the region's agricultural gross value added in one specific product *i* as a percentage of the region's total agricultural value added, and

$\left(\frac{AGVA}{TGVA}\right)_n^i$ corresponds to the share of the nation's agricultural gross value added in the same specific product *i* as a percentage of the nation's total agricultural value added.

The table shows the products for which each region of any selected country is relatively specialized. This specialization can be strong, as in the case of fruits and vegetables in the Communidad Valenciana region (ratio of 70.3 and well above the national average), or weak as in the case of wine in Galicia (ratio of 5 and close to the national average).

Table 1.7 Main agricultural regions in the EU (1999)

Regions	Major products (% of region's total)
Belgium	
Brussels gewest	Pig meat (32.7), fruits and vegetables (20.5), eggs and poultry (10.2)
Région walonne	Cattle (25), milk (23.4), other crops (16.4), cereals (11.3)
Germany	
Niedersachsen	Pig meat (25.4), other crops (9.1)
Schleswig-Holstein	Milk (31.2)
Hamburg	Fruits and vegetables (38.2)
Bremen	Fruits and vegetables (31.6), cattle (10.6)
Nordrhein-Westfalen	Pig meat (30.5), fruits and vegetables (13.1), cereals (12.1)
Rheinland-Pfalz	Wine (44.2), fruits and vegetables (11.8)
Baden-Württemberg	Fruits and vegetables (15.4)
Bayern	Milk (35.6), cattle (16.6), other crops (8.5)
Hessen	Fruits and vegetables (13.1), cereals (12.1)
Saarland	Milk (27.3), cattle (18.2), cereals (12.8)
Berlin	Fruits and vegetables (51.3)
Brandenburg	Milk (29.6), cereals (15.6), eggs and poultry (9.9)
Mecklenburg-Vorpommern	Milk (29.9), cereals (23.1), other crops (16.4), eggs and poultry (6.4)
Sachsen	Milk (32), cereals (24.6), other crops (17.5), eggs and poultry (7.3)
Thüringen	Milk (26.1), cereals (19.2), other crops (9.6), eggs and poultry (6.6)
Spain	
Galicia	Milk (26.6), cattle (18.9), other crops (11.8), eggs (10.9), wine (5.0)
Principado de Asturias	Milk (47.3), cattle (31.8)
Cantabria	Milk (56.6), cattle (29.1)
País Vasco	Milk (21.5), wine (16.9), other crops (12.8), cattle (12.1), eggs and poultry (9.3), cereals (6.1)
La Rioja	Wine (35.5), fruits and vegetables (29.9), other crops (10), cereals (6.4)
Cataluña	Pig meat (29.6), eggs and poultry (9.9), cattle (9.0)
Communidad Valenciana	Fruits and vegetables (70.3)
Región de Murcia	Fruits and vegetables (51.8), pig meat (14.1)
Andalucía	Fruits and vegetables (39.6), other crops (7.9)
Canarias	Fruits and vegetables (62.6), eggs and poultry (9.1)
Navarra	Cereals (16.8), pig meat (15.1), milk (10.1), wine (6.5)

Table 1.7 (Continued)

Regions	Major products (% of region's total)
Aragon	Pig meat (29.3), cereals (14.8), cattle (10.4), eggs and poultry (9.8)
Baleares	Fruits and vegetables (53.3), eggs and poultry (9.9), milk (9.1)
Castilla Léon	Pig meat (17.3), other crops (16), milk (13.4), cattle (13.1)
Madrid	Milk (18.3), eggs and poultry (15.9), cattle (10.8), cereals (8.0)
Castilla la Mancha	Wine (19.2), eggs and poultry (8.1), cereals (6.6)
Extremadura	Pig meat (17.3), other crops (13.8), cattle (13.3)
France	
Île-de-France	Cereals (31.2), other crops (20.7), fruits and vegetables (18)
Champagne-Ardennes	Wine (37.3), cereals (20.7), other crops (16)
Picardie	Cereals (26.2), other crops (33)
Centre	Cereals (33.3), other crops (13.6)
Haute Normandie	Milk (23.8), cereals (23.3), other crops (18.2), cattle (16.3)
Basse Normandie	Milk (42.1), cattle (23.3)
Bourgogne	Wine (32.9), cattle (18.6), cereals (17.7)
Nord Pas-de-Calais	Other crops (23.2), milk (21.9), cereals (18.7), fruits and vegetables (11.1)
Lorraine	Milk (33.6), cereals (24.9), cattle (19.9), other crops (8.0)
Alsace	Wine (33.7), cereals (20.5)
Franche-Comté	Milk (50.1), cattle (17.3)
Pays de la Loire	Milk (21.8), eggs and poultry (18.7), cattle (18.4)
Bretagne	Pig meat (30.2), milk (22.2), eggs and poultry (20.8)
Aquitaine	Wine (36.6), cattle (14.8), fruits and vegetables (12)
Limousin	Cattle (54.1)
Auvergne	Cattle (29.9), milk (29.1), cereals (14.1)
Rhône-Alpes	Milk (20.8), fruits and vegetables (16.6), wine (16), eggs and poultry (10.4)
Languedoc Roussillon	Wine (52.7), fruits and vegetables (29.1)
Provence Alpes Côte d'Azur	Fruits and vegetables (35.9), wine (32.2)
Corse	Fruits and vegetables (39.2), wine (21.8), pig meat (8.2)
Italy	
Piemonte	Cereals (20.9), cattle (17.9), wine (10.9), pig meat (7.5)
Valle d'Aosta	Milk (42.1), cattle (30.6)

Lombardia	Milk (27.1), pig meat (17.5), cattle (14.7), cereals (13.5), eggs and poultry (10.3)
Trentino-Alto Adige	Fruits and vegetables (49.1), milk (19), wine (13.9)
Veneto	Eggs and poultry (17.9), wine (12.8), cereals (11.8), cattle (11.7), other cereals (4.6)
Friuli-Venezia-Giulia	Cereals (23.4), wine (13.1), pig meat (8.8), other crops (7.7)
Emilia Romagna	Milk (15), eggs and poultry (13.3), pig meat (10.9), other crops (9.0)
Toscana	Wine (16.2), cereals (10.8)
Umbria	Cereals (16.7), other crops (15.3), pig meat (13.8), eggs and poultry (12.7)
Marche	Cereals (16.2), other crops (11.3), eggs and poultry (11.3), wine (10.5), pig meat (7.2)
Lazio	Fruits and vegetables (34.2), milk (13.1), wine (10.4), cattle (9.3)
Campania	Fruits and vegetables (40.1), other crops (9.5)
Abruzzi	Wine (27.7), fruit and vegetables (24.5)
Puglia	Fruits and vegetables (41.1), wine (13.5)
Basilicata	Fruits and vegetables (37), cereals (17.6)
Calabria	Fruits and vegetables (34)
Sicilia	Fruits and vegetables (49.7), wine (14.7)
Sardegna	Milk (26.8), fruits and vegetables (19.6), cattle (11.2), pig meat (8.6)
Molise	Eggs and poultry (16.2), cereals (14.9), milk (13.5), cattle (10.6), other crops (8.6)
United Kingdom	
Yorkshire-Humberside	Cereals (19.3), pig meat (19.2)
South-West	Milk (39.4), cattle (18.1)
East Midlands	Cereals (21.9), fruits and vegetables (14.6), eggs and poultry (14.0)
West Midlands	Milk (27.1), eggs and poultry (14.0)
Wales	Milk (40.2), cattle (25.5)
Scotland	Cattle (26.5), cereals (15.6)
Northern Ireland	Milk (32.9), cattle (30.7), pig meat (9.1)

Source: CEC (2002), T/50 to T/54.

In Germany, only four länder out of 16 have a strong agricultural orientation, that is they display a ratio above 33 per cent. The Rheinland-Pfalz region is well known for its wine, whereas Bayern is strongly specialized in the production of milk. Attempts, made in recent years, at diversifying the agricultural production structure of Germany are visible through the development of non-traditional sectors such as fruits and vegetables in the regions of Berlin and Hamburg, where they represent more than 50 and 38 per cent of the region's agricultural production respectively. In Spain, the northern regions of Castilla-León, Aragón and Galicia produce cereals, meat and milk, whereas the southern regions of Andalucia, Extremadura, Murcia are more specialized in the production of fruits and vegetables and other crops. High specialization ratios are recorded for fruits and vegetables in the Communidad Valenciana region (70.3), Canarias (62.6) as well as Baleares (53.3). Milk is a substantial agricultural activity in Cantabria (56.6), whereas the wine from La Rioja region is now well established on international markets. The diversity of the French agricultural sector, already discussed above, appears clearly through the different specialization patterns of its regions. The western regions of France (Bretagne, Basse and Haute Normandie) are relatively specialized in meat (pig meat in Brittany) as well as milk products. The production of cereals is geographically concentrated in the center part of France (wheat in Picardie, Centre and Île-de-France). With 52.7 and 29.1 per cent of its agricultural production in the wine and in the fruits and vegetables sectors respectively, the Languedoc Roussillon region is the least diversified of all French agricultural regions. Its being so heavily specialized in the wine sector renders it vulnerable, particularly in a sector where quality dictates much of consumers' behavior. Two-thirds of all wines of medium-to-low quality, produced each year in France, come from the Languedoc-Roussillon region. In the 1980s, the enlargement of the EU to the Mediterranean countries such as Greece, Spain and Portugal, has severely hit this region. A sudden over production of medium-to-low quality wine has led to a long-term policy centered on the discouragement of wine production in the non-viable wine-growing areas of the EU. The Champagne-Ardennes region remains the French wine region *par excellence*, with nearly 40 per cent of its agricultural production being generated from the wine sector. Finally, Corsica is strongly involved in the production of fruits and vegetables, a feature common to other Mediterranean regions of France (such as Provence, Alpes et Côte d'Azur) and Europe.

Most of the Italian regions recorded near the bottom of Table 1.7 belong to the less economically advanced *Mezzogiorno*. Agricultural production in these southern regions is often heavily concentrated on one product category only. For example, the production of fruits and vegetables represents between 34 and 49.7 per cent of total agricultural production in Calabria, Lazio, Basilicata, Campania, Puglia and Sicilia. A high percentage (49.1 per cent) is also observed in Trentino-Alto Adige, a mountainous region of the north where the topography places a limit to diversification. In all other agricultural regions of Italy, food production is more diversified. Finally, five out of the 12 regions of the UK have an agricultural gross value added proportionately greater than that recorded at the national level. Among these regions, two have a relatively strong meat and dairy sector: this is the case for Wales and Northern Ireland. The other three regions are relatively engaged in the production of milk (South-West and West-Midlands) as well as cattle (Scotland).

1.3 The role of agriculture in economic development

Before studying the way the two concepts are inter-related, it is worth defining what we mean by economic growth and by economic development.

1.3.1 Economic development and economic growth

Perroux (1983: 26) defines growth as 'the increase in the size of a unit, usually a country, expressed in terms of its national product in relation to the number of inhabitants'. The word 'growth' must be considered in a long period of time. Consequently, economic growth is normally understood to refer to improvements in the standard of living of a given population in a given country. Economic development normally results from high growth rates sustained over a long period of time. Although it is convenient to equate economic growth with economic development, Perroux (1983) warns against the fact that unfortunately growth can take place without development. In his own words (op. cit. p. 36):

> This danger obviously exists in developing countries when economic activity is concentrated around branches of foreign firms [...]. Even in developed countries we see that, as growth progresses, the benefits of development are being unevenly distributed in geographical terms, because relatively 'empty' regions still exist, and in social terms, because 'pockets of poverty' have not disappeared.

During the 1980s, the notion of economic development has been reviewed to incorporate one of the most visible negative and lasting effects of economic growth: environmental degradation. Because of increasing pollution, desertification, poisoning of water supplies, soil erosion and other environmental degradations in both the developed and developing countries, many have argued that for development to be acceptable it must be 'sustainable'. The World Commission on Environment and Development (1987: 43) defines sustainable development as 'development that meets the needs of the present without compromising the ability of future generations to meet their own needs.' A more complete definition of economic development would refer to increasing per capita incomes, minimization of externalities, poverty eradication and reduction in social and geographical inequalities. In the next paragraph, we will focus on agricultural development.

1.3.2 Agricultural development theories

Agricultural development theories are aimed at explaining how the basic sources of growth – labor, capital, natural resources – can be best combined and optimized to generate broad-based agricultural growth. These theories can be classified into two major groups:

(i) theories centered around the locational aspects of agricultural production, and
(ii) theories emphasizing innovation as a motor of growth.

1.3.2.1 The location and diffusion theory

Location theory in general is about the search of the firm's best location at a particular point in time, given a certain set of circumstances. In particular, it studies the intensity of agricultural production in relation to the distance from urban centers and to the nature of transportation systems. It is a widely recognized fact that agricultural development on a regional basis is a function of the proximity to urban markets and of available transportation systems. The distance from cities and the availability of transport networks do matter because of differences in transport and marketing costs, as well as in the ease of obtaining more productive inputs and quality services.

Von Thünen's work (1826) on the location and optimal intensity of farm enterprises in relation to urban areas, was to set the founding pillars of the location theory. His work was subsequently extended by geographers and later, by industrial economists interested in the spatial location and distribution of manufacturing activities.[7] In particular, Weber's contribution published first in 1909 is considered as the seminal article dealing with the location of manufacturing firms (Weber, 1909). He developed a location theory based on a least-cost approach, which greatly emphasized transport costs. An interesting contribution is also the one by German economist Friedrich List (1841) who advocates the association between manufacturing and agricultural sectors on the same territory to allow the diffusion of knowledge facilitated by spatial proximity.

Von Thünen developed the model of 'the dual economy' represented by a large town in a very fertile plain surrounded by 'wilderness'. His major contribution is his discovery of the marginal productivity theory of distribution. According to Nerlove (1988), Von Thünen's model can be extended into a model of 'dual economy development' which is very applicable to the case of developing countries where high transport costs are a substantial determinant of relative commodity prices. Finally, Samuelson (1983) formalized Von Thünen's model and developed it into a general spatial equilibrium model.

The location theory as applied to agricultural development inevitably leads us to the notion of linkages. The diffusion theory stresses the importance of backward and forward linkages involving tangible (factor inputs) and intangible (services, information) assets among farmers and between farmers and other units situated in a given geographical area. Of particular relevance is the transfer of information from the most advanced farmer to the laggards, and of critical importance is the diffusion of an innovation as variables promoting productivity growth. Diffusion aspects have provided the grounding for the development of extension services. As thoroughly documented by Von Hippel (1988), synergies between manufacturing firms facilitate the process of technical innovation.

1.3.2.2 Technology-related theories

The theory of induced innovation states that technical change in agriculture is a response to changes in resource endowments, as well as to growth in product demand. Factor endowments refer to the relative abundance of factors of production in a given country. 'Natural' factor endowments specifically refer to natural resources (climate, soil and minerals) as well as to unskilled labor, whereas 'acquired' factor endowments suggest the use of capital, technology and skilled labor. The idea that a variation in relative factor prices can explain technical change was first advocated by the classical economists during the 19th century, such as David Ricardo (1817) for example, and it was applied to the field of agricultural economics in the 1960s with the Hayami and Ruttan model (see Chapter 5). The demand-induced technical change hypothesis rests on the observation that markets with a relatively high demand are taken as a signal that profitable opportunities for investment in innovation exist, whereas in depressed markets, sluggish demand creates a pressure for firms to innovate and to invest (Schmookler, 1966; Fransman, 1986).

The first observation rests partially on the hypothesis that demand for the new product increases with the increase in income and with the consumer's understanding of the product. The next section studies the causal link existing between income changes and agricultural activity.

1.3.3 Trends in agricultural activity

Agricultural activity is a declining portion of the economies of all countries in the world. The importance of agricultural production and employment declines with economic development, whereas the demand for manufactured goods and services expands. This basic law has been observed in every country. For the individual EU countries, the declining importance of agriculture, in terms of both employment and contribution to GDP, is given in Table 1.8.

Movement out of agriculture is very noticeable in every country. In less than 40 years, agricultural employment has been reduced by more than three-quarters in Europe. The decline has been most marked

Table 1.8 Share of agriculture in total employment and GDP at EU level (1960 and 1999)

	Employment in agriculture*		Share of agriculture in GDP – 1999
	1960	**1999**	
EU-15	21.1	4.5	1.8
Austria	18.7#	6.2	1.2
Belgium	8.7	2.4	1.2
Denmark	18.2	3.3	2.0
Finland	24.4#	6.4	0.9
France	22.5	4.3	2.4
Germany	13.8	2.9	0.9
Greece	57.1	17.0	7.1
Italy	32.6	5.4	2.6
Ireland	37.3	8.6	2.9
Luxembourg	16.6	2.0	0.7
The Netherlands	9.8	3.2	2.4
Portugal	43.9	12.6	3.3
Spain	42.3	7.4	4.1
Sweden	8.1#	3.0	0.7
UK	18.4	1.6	0.9

* As a percentage of total civilian employment. Are also included numbers employed in the forestry, hunting and fishing sector.
\# Data for 1970.
Source: CEC (2002), Tables T/27 and T/118.

in the UK where the agricultural labor force has been reduced nearly by 90 per cent over the period under review. Greece has been much slower to adapt, and it still has the biggest agricultural workforce in the EU. Another striking feature emanating from the table is the uneven contribution made by agriculture in the various EU countries. By European and OECD standards, Greece (with 7.1 per cent of total GDP represented by agriculture) and Spain (4.1 per cent) still rely relatively substantially on their agricultural sector. In the most advanced European economies, this share is normally well below 2.0 per cent. Italy (2.6), France (2.4 per cent) and The Netherlands (2.4 per cent) deviate from this simple rule due to their multi-century tradition in agriculture, to their internationally recognized know-how and quality (France and Italy), and to their efficient production and commercialization system (The Netherlands). In value terms, Holland is the second largest exporter of agricultural products after the USA, a ranking which is sometimes contested by France.

On a much wider scale, and when taking into account minor deviations from the rule, there seems to be a negative association between per capita national income, the proportion of national income derived from agriculture and the proportion of the labor force in agriculture. At one particular point in time, most of the production, employment and consumption of a low-income country involves food, as reflected in Table 1.9(a).

As can be seen, richer countries such as Germany and Italy in the EU, and the USA and Japan elsewhere, devote normally an extremely limited share of their labor resource to the production of food (between 2.7 and 5.8 per cent of total employment). Concomitantly, agriculture represents but a very low share of total GDP (that is, less than 2.7 per cent). By contrast, poorer countries such as Ethiopia allocate a large amount of their labor resource to the agricultural sector, which represents more than half of domestic GDP.

In introducing a dynamic stance in the analysis, we see that countries have been adjusting gradually to the new international division of labor as well as to new emerging demand patterns (Table 1.9(b)).

Tables 1.9(a) and (b) seem to suggest the existence of a simple rule: the less important the agricultural sector, the more 'developed' the country. Moreover, economic development typically proceeds by stages. Rostow (1960a,b) first highlighted this fact, and the idea was later expanded by Balassa (1977) in what became known as 'the stages approach to comparative advantages'. According to this approach, a country's comparative advantage evolves in stages. In the first stage of development, a country normally starts intensifying its production of

Table 1.9(a) Relationship between per capita national income and the importance of the agricultural sector in the economy. Selected countries (1999)

	GDP per capita in US$*	Agricultural VA/total GDP (%)	Agricultural labor force/ total labor force in 1998 (%)
Mozambique	861	33.0	–
Ethiopia	628	52.0	88.6~
Bangladesh	1,480	25.0	63.2~
India	2,250	28.0	66.7~
Kenya	1,020	23.0	–
China	3,620	17.0	48.6~
Senegal	1,420	18.0	–
Ecuador	2,990	12.0	7.3
Colombia	5,750	13.0	10.0
Thailand	6,130	10.0	51.3
Turkey	6,380	16.0	43.4
Brazil	7,040	8.6	23.7
Hungary	11,400	5.7#	7.5
Greece	15,400	7.4#	17.8
Spain	18,100	3.8	8.0
Italy	22,200	2.7	5.8
Germany	23,700	1.1	2.9
USA	31,900	–	2.7
Japan	24,900	1.7#	5.3

* Current international $ in PPP (purchasing power parity).
# Data for 1998.
~ Data for 1995.
Source: World Bank (2001).

agricultural commodities and raw materials. Indeed, at the beginning, the primary needs to be fulfilled are food and export earnings in the case of an open economy. Food is a necessity good, that is, a commodity for which the income elasticity of demand is less than unity. The production of non-processed food products and the extraction of raw materials require minimum capital and skills, but abundance of labor and natural resources. The exploitation of natural factor endowments in a given country allows the agricultural sector (*stricto sensu*) and/or the primary products sector to become the backbones of economic development. As development and growth proceed and as international competition intensifies, the country moves towards labor-intensive manufacturing activities, if labor is a relatively abundant factor (stage 2). Eventually the country starts building up its own food industry. During this second

Table 1.9(b) Sectoral share of GDP (Selected countries, 1960 and 1992) in (%)

	1960	1992
Agriculture		
China	39*	27
India	50	32
Low income economies**	49	29
Upper middle income economies	28	10 (1986)
High income economies of which: Sweden	6	na
	7	2
*Industry****		
China	38*	34
India	20	28
Low income economies**	12	31
Upper middle-income economies	24	40 (1986)
High income economies of which: Sweden	40	na
	40	32
Services		
China	23*	38
India	30	40
Low income economies	39	40
Upper middle-income economies	48	50 (1986)
High income economies of which: Sweden	54	na
	53	66

* Figures for 1965.
** This excludes China and India.
*** Industry here includes manufacturing, building construction and extractive activities.
Source: World Bank. World Development Report, Various Issues (New York: Oxford University Press).

stage of development, a typical developing country will derive comparative advantages in labor-intensive industries, such as clothing-textiles. With rising living standards and wages, these labor-intensive activities will be phased out, and the country will develop capital and subsequently skill-intensive activities. This is the third stage, a stage during which capital-intensive manufacturing activities, such as steel, shipbuilding and motor vehicles, emerge. Finally in the last stage, high value-added manufacturing activities, using skilled labor and state of the art technologies become a prerequisite for sustained growth. This stage is also characterized by the dominant contribution of services and of information technology-related activities (semi-conductors, telecommunications and information processing) to economic growth. The movement from one stage to the next coincides with increasing living standards. A typical country, that is, most European countries, the USA, South East Asian Countries, Japan and China, evolves along this path.[8] This pattern of development is clearly reflected in the export structure of a country.

Over a certain period of time, economies adjust structurally away from agriculture (and primary products) towards labor-intensive and skill or knowledge-intensive manufacturing and service activities. The rate of adjustment varies across countries. Undoubtedly, countries experiencing higher growth rates adjust more rapidly. Table 1.9(b) shows clearly a faster rate of adjustment of China compared with India.

Agriculture is only a starting point in the developmental trajectory of a country. Why does the share of agricultural output and employment decrease ineluctably? An insight into some basic principles and aspects of international trade theory may provide an explanation.

1.3.4 Explaining the movement away from agriculture

The modality in which agriculture gradually recedes in favor of other economic sectors is depicted in Figures 1.1(a) and (b) which indicate the equilibrium between production and consumption in a closed economy. In Figure 1.1(a), AA represents *the production possibility curve* which describes the country's production or supply conditions. Working under the assumption of increasing opportunity costs results in a production possibility curve which is concave to the origin.[9] Thus, for each additional unit of all other goods produced, more agriculture has to be sacrificed. In the same figure, II represents the *community indifference curve* which indicates the different combinations of two commodities (x, y) that yield the same level of satisfaction to the community or country. In autarky, equilibrium is reached at point E, where the nation's production possibility curve is tangent to the community indifference curve. At that point, supply equals demand for both agricultural commodities and all other goods.

Factor expansion or productivity growth pushes the production possibility curve outwards to the right, which becomes A′A′. Assume first that this takes place in an equi-proportional manner, so that the outputs of the two goods increase in the same proportion. Similarly, we can assume that consumption of agricultural commodities and other products increases by the same proportion, resulting in the new community indifference curve I′I′. In this case, the new equilibrium between production and consumption sets at E′. We can see that because of the assumption of equi-proportionality, growth had a *neutral effect* on both production and consumption, in that the proportionate

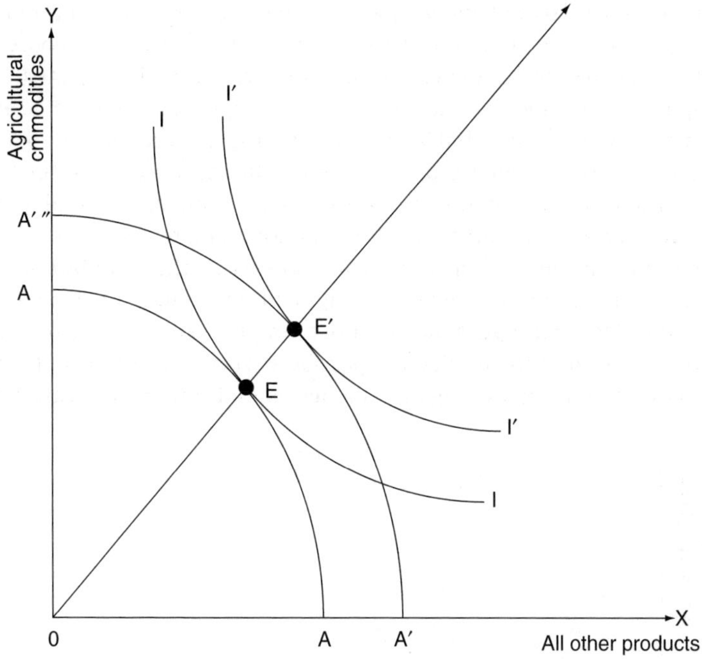

Figure 1.1(a) Autarky equilibrium and neutral effects of growth

increase in consumption of each of the two commodities is exactly matched by the proportionate increase in their outputs.

However, equi-proportionality rarely holds in practice. Indeed, a fundamental fact observed by agricultural economists, and as we will see more thoroughly in Chapter 2, is that income elasticity of demand for food is normally less than unity, that is:

$$\varepsilon_y = \frac{\Delta Q}{\Delta Y} \cdot \frac{Y}{Q} < 1$$

First perceived by the German statistician Ernst Engel in the 19th century, this simple rule has become the most widely established and empirically tested law in economics. A myriad of studies on patterns of consumption in different countries have concluded that if prices and demographic variables (such as the dimension and the composition of the family unit) remain constant, an increase in income will lead to a less than proportionate increase in the consumption of food. Eventually after a critical level of income, a level that enables food commodities not to be considered as necessities any more, an increase in income will not be matched by an increase in the quantity of food consumed. This simple but fundamental rule has important implications for the farming community. It means that as economic growth proceeds, the rise in demand following such growth patterns affects increasingly the non-agricultural sectors of the economy. As a result, agricultural prices, which assume their market clearing function, tend to decline relative to other prices in the economy. As growth proceeds, the equilibrium price of all other commodities relative to agricultural commodities increases. This translates through a steeper price line P′P′ compared with PP in Figure 1.1(b), after the production possibility curve has shifted outwardly and equi-proportionately again. The new equilibrium E′ in Figure 1.1(b)

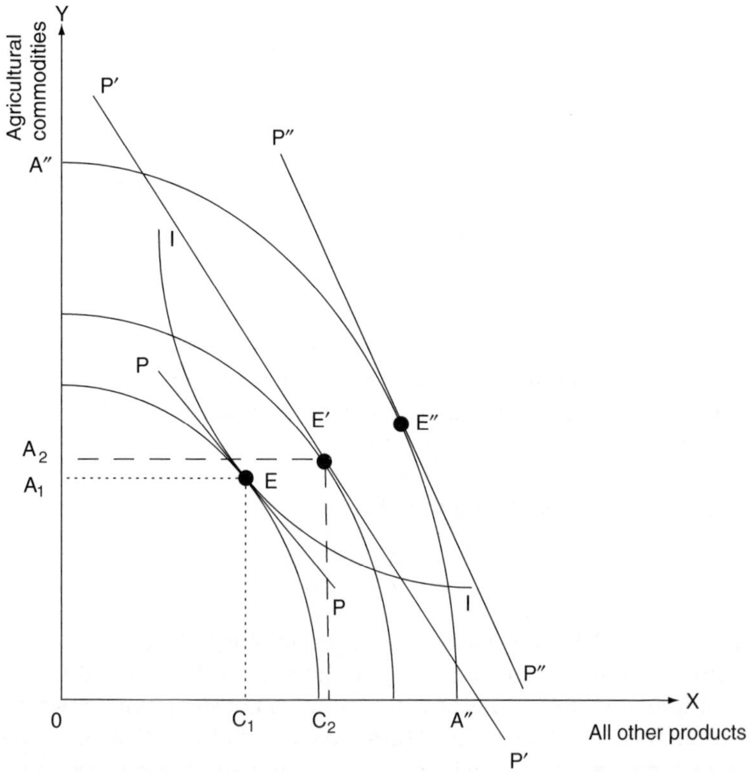

Figure 1.1(b) Effects of growth with income elasticity of demand for food less than unity (closed economy)

shows that even though the quantity of food consumed and produced has risen from A_1 to A_2, the quantity of all other goods consumed and produced has risen by a much larger proportion from C_1 to C_2.

Agriculture's share in national product would fall even more if we include in our model a second observed fact which is that productivity growth is relatively faster in agriculture than in other sectors of the economy (see Chapter 5). This would allow the production possibility curve to assume a new shape as represented by A″A″ in Figure 1.1(b). The new equilibrium point E″ shows how economic growth typically results in a marginal increase in the quantity (demanded and supplied) of agricultural commodities and in a substantial increase in the quantity of all other commodities.

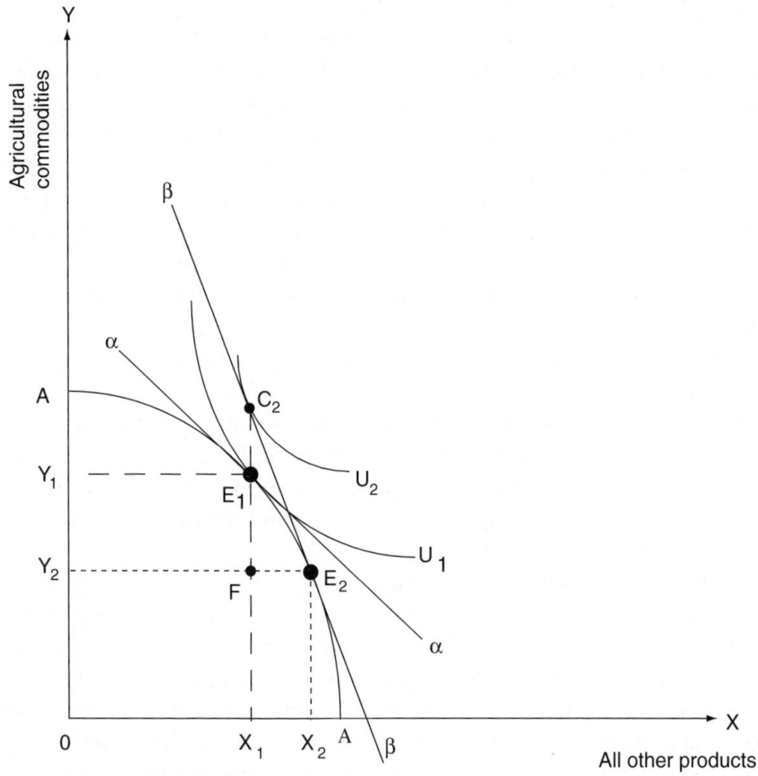

Figure 1.2(a) Changes in consumption and production patterns arising from free trade

Going one step further allows us to study the equivalent equilibrium conditions in the framework of an open economy. Figure 1.2(a) shows the same production and consumption conditions for an open economy. The assumptions underlying the model represented in Figure 1.2(a) are that:

(i) all products are tradable internationally,
(ii) the curves depict the production and consumption conditions in an agrarian and small economy,
(iii) the rest of the world is more industrially advanced.

What is meant by 'small' is the fact that the country is a price taker on international markets.

The $\alpha\alpha$ line represents the price ratio had the small agrarian economy (SAE) chosen not to trade. Free trade allows *specialization* and leads eventually to *factor price equalization*. As we will see in Chapter 7, specialization is implied by Smith's and Ricardo's theories (although they strictly mean *inter-industry* specialization), and factor price equalization is deduced from the Heckscher-Ohlin model. Samuelson's factor price equalization theorem can be briefly summarized as follows: each country exports the commodity that uses the relatively cheap (and abundant) factor in autarky; higher demand for that commodity leads to an increase in its price; the rising (domestic) demand for the abundant factor causes its price to rise. Conversely, declining demand for the scarce (and expensive) factor causes its price to decline. As a result, free trade leads to equalization of relative factor prices for all trading nations.

At the global level, factor price equalization occurs when the price ratio lines, of the SAE and of the rest of the world, are confounded. This relative cost equalization occurs at $\beta\beta$. The price ratio line $\beta\beta$ is the international price ratio. $\beta\beta$ is also called the terms of trade (ToT) of the small agrarian country under review.[10]

For the SAE, because the price ratio changes (from price ratio lines $\alpha\alpha$ to $\beta\beta$) as a result of free trade, the indifference and production possibility curves no longer need to be tangent. The excess of demand and supply will be matched by corresponding imports and exports. Because line $\beta\beta$ is steeper than $\alpha\alpha$, agricultural commodities become cheaper in the small agrarian country, relative to other commodities. Under free trade, general economic growth elsewhere, higher productivity in the agricultural sector as well as lower prices prevailing in the most advanced nations of the rest of the world are diffused until they set in motion an adjustment process in the agrarian and less economically advanced country. A steeper price ratio line means in our case that agriculture's terms of trade decline, or in other words that more exports of agricultural products are needed in order to buy the same amount of imports of non-agricultural products.

The new price ratio line $\beta\beta$ is now tangent to the production possibility curve AA as well as to a new indifference curve U_2 corresponding to a higher level of satisfaction. This implies that free trade leads to a higher level of welfare. The production point shifts from E_1 to E_2. This new equilibrium point shows that the production of non-agricultural

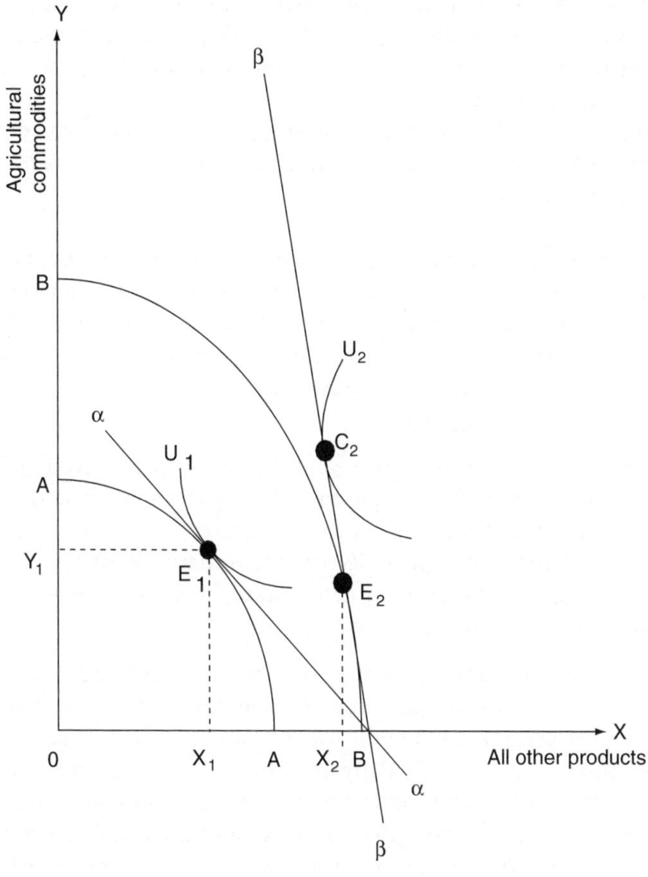

Figure 1.2(b) Growth effects in an open economy

commodities will expand, whereas the production of agricultural goods will be curtailed (from Y_1 to Y_2). The country can import C_2F agricultural products from the more efficient economies of the rest of the world and can export FE_2 other commodities.

Finally, Figure 1.2(b) illustrates the same situation as in Figure 1.2(a) but it takes into account growth effects as well. The shape of the new production possibility curve BB shows the rapid productivity effects in the agricultural sector. The new international price ratio $\beta\beta$ is now tangent to point C_2 on the new indifference curve U_2 yielding a higher level of satisfaction. The shift from E_1 to C_2 shows that the consumption of both agricultural and other commodities has increased, but that the increase in the consumption of agricultural commodities has been less than that of all other goods (Engel's law). The new production point E_2 suggests that less agricultural products will be produced by the country, whereas the production of all other commodities will expand substantially (from X_1 to X_2).

When growth effects as well as the Engel's law are taken into account, the similarities and differences between autarky and free trade can be summarized as follows:

(1) In both situations, the production and consumption of agricultural products expand proportionately less than that of all other commodities; in some cases, the production and consumption of agricultural products decrease. In line with the Rostow/Balassa 'stages approach' to development and comparative advantage, the developing economies become less and less agrarian.

(2) In both cases, the new price ratio line is steeper. According to the diagrams above, this means that agriculture's terms of trade decline in international markets. The farming community of a growing economy is faced with ineluctable price decreases, no matter what type of international trade policy is chosen, with autarky and total free trade being the two extremes of the same spectrum.

(3) If the type of trade policy does not reverse the trend of declining farm prices in a growing and relatively agrarian economy, free trade will however accentuate the adjustment process, when compared with protectionism. This is essentially true for an SOE that benefits from overall growth effects arising in the rest of the world, as well as from productivity effects in the agricultural sector of the rest of the world. In a free trade situation, the terms of trade of agriculture decline faster, and adjustment towards a more efficient agricultural sector is faster.

Summary

Agriculture first became a distinctive field of economic analysis with the work of French physician and physiocrat François Quesnay in 1758. Modern visions of the agricultural system visualize the area of agriculture as a set of interrelated activities encompassing farming, processing of food products, research and public sector based policies. Given the importance of the agri-business sector in EU countries, the notion of *filière* as developed by the French economists is also of importance to modern nations.

The increase in world food production in the last three decades or so, has by far exceeded the rise in the population growth rate, disproving British moral philosopher Malthus' thesis. As a result, the world agricultural system produces enough food for every individual on earth. Although the western economies have had to reorder their agricultural system by curbing oversupply in some specific sectors, Asian countries have experienced a large increase in food supply during the 1990s. Cereals and milk are the two major broad products produced at the global level. For the EU, the production of meat products represents nearly 20 per cent of total production, whereas fruits and vegetables are another important commodity. EU countries show a contrasting picture in terms of specialization, with the North European members being more engaged in the production of dairy products, meat and cereals. However, an effort in terms of diversification in large countries such as Germany should be noted, where some Nordic regions (that is, Hamburg) have intensified the production of Mediterranean products in the last decade.

The study of the association between the importance of agriculture as an economic activity and the relative wealth of a nation tends to point towards an inverse relationship between the two phenomena. Richer countries in the world are also those that devote a smaller proportion of their resources to the production of agricultural products. This association is best summarized by Rostow-Balassa 'stages approach to economic development'. Explanations for the gradual retreat of agriculture in favor of other manufacturing and service activities are provided by productivity effects (on the supply side) and by Engel's law (on the demand side).

Key terms and concepts

Agricultural system
Agri-food 'filière'
World food supply

Revealed comparative advantage
Utilized agricultural area (UAA)
Agricultural regions
Factor endowments
Economic growth and agriculture
Factor price equalization
Production possibility curve
Community indifference curve

Notes

1 The word 'physiocrat' stems from the Greek φυσις (nature) and κρατος (power). Note that the physiocrats pioneered the concept of *laissez-faire, laissez-passer* ('let it be; let it go').
2 All figures in this paragraph are from CEC (2002).
3 All statistics in this section are, unless otherwise specified, from FAO (2001): Yearbook Production, Volume 53.
4 The *ceteris paribus* clause implies here that cataclysms (that is, wars, epidemics, and so on) are excluded from the analysis.
5 As for endnote 2.
6 The number of people working on the farm was actually well above 10 million people, but the figure of 6.9 million is based on full-time equivalent workers.
7 The interest by industrial economists in spatial aspects of manufacturing activities is relatively recent, in spite of Hotelling's early contribution (see Hotelling, 1929).
8 Obviously, there are exceptions to this development path: Ireland is one example in the case of the EU.
9 Indeed, in the real world, a commodity's opportunity cost goes up as its production increases; increasing costs result from the fact that factors of production are imperfect substitutes one to another.
10 The terms of trade of a country are the relative prices at which two commodities are traded in the international market. The ToT describe the relationship between the prices a nation pays for its imports and the prices it receives for its exports.

Bibliography

Angelier, J.-P. (1991) *Economie Industrielle – Eléments de méthode*. Grenoble: Presses Universitaires de Grenoble.

Aujac, H. (1960) 'La Hierarchie des Industries dans un Tableau des Echanges Interindustriels', *Revue Economique*, No. 2.

Balassa, B. (1977) *The Stages Approach to Comparative Advantages* (Washington: World Bank Staff Working Papers, No. 256).

CEC (1995) *The Agricultural Situation in the European Union* (1994 Report, Luxembourg).

CEC (2002) *The Agricultural Situation in the European Union* (2000 Report, Luxembourg).

Clark, C. (1940) *The Conditions of Economic Progress* (London: MacMillan).

de Bernis, G. (1966) 'Industries Industrialisantes et Contenu d'une Politique d'Intégration Régionale', *Economie Appliquée*, No. 3/4.

FAO (1991) *Yearbook Production*, Volume 45, Food Agricultural Organization: Rome.

FAO (2001) *Yearbook Production*, Volume 53, Food and Agriculture Organization of the UN: Rome.

FAO (2002) *The State of Food Insecurity in the World 2002*. FAO Report, October, Rome.

Fisher, A.G.B. (1939) *The Clash of Progress and Security* (London: MacMillan).

Fourier, C. (1822) *Traité de l'association domestique agricole* (Paris: Bossange et Cie).

Fransman, M. (1986) *Technology and Economic Development* (Brighton, Sussex: Wheatsheaf Books Ltd).

Hotelling, H. (1929) 'Stability in competition', *Economic Journal*, 39: 153, March, pp. 41–57.

List, F. (1841) *National System of Political Economy* (London: Dent).

Leontief, W. (1951) *The Structure of the American economy 1919–1939*, 2nd edn (Oxford and New York: Oxford University Press).

Malthus, T.R. (1803) *An Essay on the Principle of Population*. Penguin Books Limited, 1988 edition.

Montfort, J. (1983) 'A la recherche des filières de production', *Economie et Statistique*, No. 151, January, pp. 3–12.

Nerlove, M. (1988) Von Thünen's model of the dual economy (Proceedings of the XXth International Conference of Agricultural Economists. *Agriculture and Governments in an Interdependent World*, August 1988, Aldershot: Dartmouth).

Norton George, W. and Alwang, J. (1993) *Introduction to Economics of Agricultural Development* (London and New York: MacGraw-Hill).

Perroux, F. (1973) 'L'effet d'entraînement: de l'analyse au repérage quantitatif', *Economie Appliquée*, No. 2/4.

Perroux, F. (1983) *A New Concept of Development* (London: Croom Helm).

Ricardo, D. (1817) *On the Principles of Political Economy and Taxation*. Reprinted in P. Sraffa and M. Dobb (eds) (1953) *The Works and correspondence of David Ricardo*, Volume I (Cambridge: Cambridge University Press).

Rostow, W.W. (1960a) *The Process of Economic Growth* (Oxford: Clarendon Press).

Rostow, W.W. (1960b) *The Stages of Economic Growth: a non-communist manifesto* (London: Cambridge University Press).

Samuelson, P.A. (1983) 'Thünen at Two-Hundred', *Journal of Economic Literature*, 21, pp. 1468–88.

Schmookler, J. (1966) *Invention and Economic Growth* (Cambridge, MA: Harvard University Press).

Seitz, W.D., Nelson, G.C. and Halcrow, H.G. (1994) *Economics of Resources, Agriculture and Food* (London: Mac Graw Hill).

Toledano, J. (1978) 'A Propos des Filières Industrielles', *Revue d'Economie Industrielle*, No. 6.

Von Hippel, E. (1988) *The Sources of Innovation* (Oxford: Oxford University Press).

Von Thünen, J.H. (1826) *Der Isolierte Staat* (Gustav Fischer, Iena edition).

Weber, A. (1909) *Über der Standort der Industrien* (English translation: Theory of the Location of Industries, 1929, University of Chicago).
World Bank (various issues) *World Development Report* (New York: Oxford University Press).
World Bank (2001) *World Development Indicators* (Washington).
World Commission on Environment and Development (1987) *Our Common Future* (New York: Oxford University Press).

2

Demand and Supply – Market Mechanisms for Agricultural Products

Objectives of this chapter

- To explain the mechanism of demand and supply interactions on agricultural markets and to highlight the specificity of demand and supply in the case of agricultural products.
- To provide a brief analysis of agricultural prices and to discuss the relationship between prices and incomes.
- To describe the structure of the agricultural sector in the EU.
- To draw attention to the 'farm problem'.

Introductory remarks

From a demand and supply perspective, agricultural products fall into three broad groups: food, raw materials and non-food consumer goods such as cut flowers, trees and shrubs. Raw materials such as cotton are of very limited importance to the EU, whereas public policy is normally extremely limited with regard to non-food consumer products. This explains why most of the attention of European agricultural economists has traditionally been confined to food products.

The concepts of demand and supply can be applied both at the level of an individual unit (the consumer or the firm) and at the level of the industry or market. As every market comprises a large number of individual consumers interacting with suppliers, the market demand curve is the sum of all quantities demanded by individual consumers, for any price level.

2.1 The demand function – price and income elasticities of demand

Knowledge about the characteristics of the market demand for a particular agricultural product is useful to producers as it helps them determine their supply schedules. Policy makers use the information relating to the market demand so as to exert an influence on the market. This can be done for example by subsidizing the consumption of a popular product or by penalizing that of a product detrimental to the health.

2.1.1 The demand function

The demand function relates the quantity demanded for a product and its different influencing variables, which are of an economic nature (prices) as well as of a non-economic type (tastes or structure of the population). In the simplest case, the demand function relates the quantity demanded to the price of a particular product. This simple relationship is best represented by a downward sloping curve connecting all the points of the *demand schedule*. These points show the quantities demanded at each price, *ceteris paribus*, that is when all the other influencing variables do not change. For normal goods, that is, in the case of most agricultural products, the demand curve can be represented as in Figure 2.1.

The demand curve in Figure 2.1 denotes that a typical consumer would tend to demand more at lower prices. This implies that demand curves normally have a negative slope,[1] and this is known as the 'law of demand'. As can be seen, a price decrease from €45 to €35 leads, *ceteris*

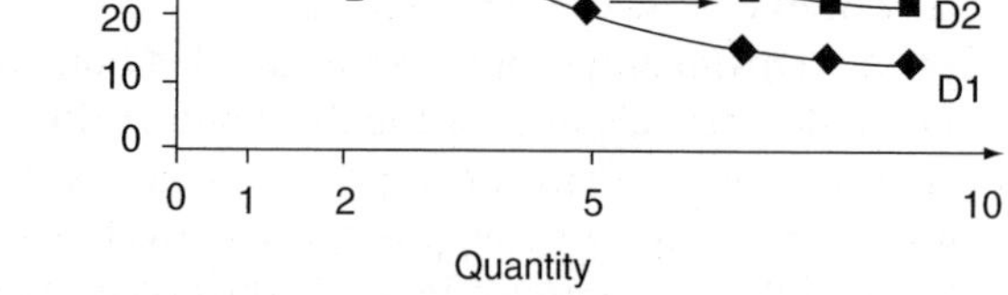

Figure 2.1 The demand curve for beef of a household

paribus, to a doubling in quantity demanded; this can be visualized by the movement along the demand curve from point A to point B.

When any variable that influences demand other than price changes, then the demand curve changes. In Figure 2.1, this is for example illustrated by a rightward shift of the demand curve from D_1 to D_2, implying that the household's quantity demanded is greater at every price level. The reasons explaining why the demand has now increased are multifold. In particular, this can be attributed to an increase in the size of the household or to a rise in the household's disposable income.

At the macroeconomic level, the demand expressed by a large number of individual consumers is called 'aggregate demand'. The factors affecting the aggregate or market demand for a particular product can therefore be expressed by the following demand function:

$$Q = f(P, P_{1\ldots n}, Y, G, N, T, A) \tag{2.1}$$

where:

Q is the quantity demanded per time period,
P is the price of the product,
$P_{1\ldots n}$ refers to the price of n other competing products,
Y is the average income per head of the population,
G can refer to the distribution of income,
N is the population,
T refers to the tastes and preferences of the population, and
A denotes the differentiating attributes of the product enhanced by advertising and technological change.

The influence that the price of other n competing products can have on the demand for a specific product is captured by the notions of *substitutable* and *complementary* products. A substitute is a good that can be used in lieu of another product, such as in the case of butter and margarine. Two products are complement if they are consumed together. Variable N can be expanded further to include data on the composition of the population. An aging population in the EU explains indeed the shift away from liquid milk consumption to soft drinks and alcohol. The tastes and preferences of the population (variable T) can in turn be explained by sociological factors such as the belonging to a specific ethnic group, the attitude towards nutrition, health and food safety. Finally, variable A has grown in importance over the last decades. It encompasses all the attributes of the product that allow the consumer to differentiate it clearly from other products. The aim of advertising is to increase the perceived (if not real) degree of differentiation between products. The increase in product differentiation is normally associated with the increase in the incidence of technological change over time.[2] New technologies in both food processing and food preparation have enabled the gradual emergence and popularity of convenience foods, and particularly of frozen and ready-to-eat products. Although the non-price and sociological variables play an increasing role in influencing demand for food products, price and income are still two major variables explaining the variations in market demand, as well as the supply and policy orientations in agricultural markets. Understanding the degree to which demand is sensitive to price and to income changes is crucial for a thorough analysis of the 'farm problem'.

2.1.2 Price elasticity of demand

The price elasticity of demand, denoted by ε_p, is the ratio of a change in the quantity demanded of a good to a change in its price, *ceteris paribus*. Expressed as the percentage change in the quantity demanded as a response to a percentage change in its price, all other things remaining unchanged, price elasticity of demand is given by the following formula:

$$\varepsilon_p = \frac{\Delta Q/Q}{\Delta P/P} = \frac{\Delta Q}{\Delta P} \cdot \frac{P}{Q} \tag{2.2}$$

with $\Delta Q = Q_1 - Q_0$ denoting the increase in output between two time periods t_1 and t_0. Consequently, Q_0, P_0, Q_1, and P_1 will refer to the quantity and price observable at time periods 0 and 1 respectively. In order to avoid selecting arbitrarily between either Q_1 or Q_0 as a reference value for the calculation of ε_p, a convention is to take the midpoint between the two values, that is, the average of the two quantities in t_1 and t_0. The same rule applies to prices. As a result, price elasticity of demand is expressed as:

$$\varepsilon_p = \frac{\dfrac{(Q_1 - Q_0)}{(Q_1 + Q_0)/2}}{\dfrac{(P_1 - P_0)}{(P_1 + P_0)/2}} = \frac{(Q_1 - Q_0)}{(P_1 - P_0)} \cdot \frac{(P_1 + P_0)}{(Q_1 + Q_0)} \tag{2.3}$$

This is known as 'arc' elasticity. In general, the most price elastic commodities are those food products that are relative luxuries with close substitutes, such as beef, fruits and vegetables. Other products such as milk and potatoes have no close substitutes and are characterized by a low price elasticity of demand. Consequently, the demand for a product tends to be more elastic the more numerous are the substitutes for it. In developing countries, the price elasticity of demand for food tends to be extremely low, particularly in the case of an increase in price. Since most of the calorie intake in low-income countries originates already from cheap starchy staple foods such as cereals and root crops, consumers have indeed little scope for shifting to less expensive foods in order to offset a rise in food prices. Also, in the developing countries, rises in food prices would have a stronger inflationary impact than in developed economies, given the predominant role of the agricultural sector in their economy.

In the EU, the price elasticity of demand for agricultural products is generally below 1 in absolute value. This implies that a price decrease will, in general have only but a small positive impact on demand.

2.1.3 Income elasticity of demand

In the same vein, the sensitiveness of demand to income changes can be computed using a similar formula. To understand the relationship existing between income and demand for food products, let's take the case of an individual i_1 with a relatively low level of income, so that his desired and optimal level of food consumption is not reached. Any increase in income is therefore likely to be spent on extra food consumption. On another hand, an individual i_2 with a relatively high level of income is likely to have reached an optimal level of food consumption. Compared with i_1, consumer i_2 spends a much smaller percentage of income on food. Constrained by physical capacity, individual i_2 may at best increase his food expenditure by switching to better quality products.[3] However, there will be a point in time at which a further increase in i_2's income will lead to a less than proportional increase in his expenditure on food. As a result, the increase in the demand for food in developed countries tends not to keep pace with the increase in consumers' incomes. This relationship between income and demand for food products was first stated by 19th century German statistician Ernst Engel. Known as Engel's law, this simple relationship stipulates that food expenditure, as a proportion of total expenditure, declines as income increases *ceteris paribus*, or that the income elasticity of demand for food tends to decline as a country moves along the development path. This is expressed as follows:

$$\varepsilon_y = \frac{\Delta Q/Q}{\Delta Y/Y} \tag{2.4}$$

Not surprisingly, income elasticity of demand for food in general is higher in the developing countries (around 0.6), compared with the developed economies, where it is around the 0.2 value (Johnston, 1991).

In the EU and OECD countries, where average household incomes are relatively high, income elasticities of demand for food products are relatively low, that is typically less than 1. Although income elasticity of demand is still positive for most agricultural products, it tends however to be negative ($\varepsilon_y < 0$) in the case of some inferior goods, that are discarded and replaced by better quality substitutes. In general, the highest coefficients (that is, close to 1 and positive) are shown for meat, fish, fruits and vegetables, as well as for diet foods. These are products that consumers tend to substitute for goods such as bread, potatoes, rice, and so on.

In 1999, the share of total household consumption spent on food was between 8.7 per cent in the UK and 14.5 per cent in Spain (EUROSTAT, 2002).

2.1.4 Concluding remarks

For most agricultural products, the price elasticity of demand is typically less than 1, in absolute value. Low price and income elasticities of demand for food products explain in great part the EU agricultural industry's relative decline with economic progress. It follows that, *ceteris paribus*, an increase in supply and an increase in sales, will lead to a decrease in total revenue, for farmers.

2.2 Characteristics of supply

Agricultural production is represented by the process of converting certain inputs – or factors of production – into a final product, or output. For our purpose, the farm is the unit of production. It takes decisions concerning the use of limited resources so as to produce agricultural products. The output of the farm can be in a form suitable for direct consumption by households, or it may require further conversion. In the latter case, the farm's output is used as an input by a firm, typically the food industry, which processes it into a consumable form.

Land, labor and capital are the farm's inputs. Capital encompasses machinery, buildings, tools, fertilizers, as well as 'human capital', that is management skills or entrepreneurship (see Chapter 5). Entrepreneurship involves taking risks and making decisions about the use of the other factors.

It is customary to distinguish between fixed and variable factors of production. This distinction refers to the time period within which it is possible to alter the quantity used of an input. In agricultural economics, the crop year (or the life cycle in the case of animals) helps the economist distinguish between short and long runs. During a crop year, or in the short run, it will be impossible for the farmer to change certain factors of production such as the surface of the land and also the capital (that is, buildings) used. These factors of production will be referred to as 'fixed'. In the short run, the farmer will nevertheless be able to change the amount of fertilizer, of fuel and of labor used. These are all referred to as 'variable' factors of production.

2.2.1 Input–output relationship – the law of diminishing returns

The input–output relationship is known as the total physical product curve shown in Figure 2.2. This curve shows that total physical product will initially rise at an increasing rate, that is, adding extra units of inputs will first of all increase output more than proportionately. Subsequently, the rate of increase in total physical product will

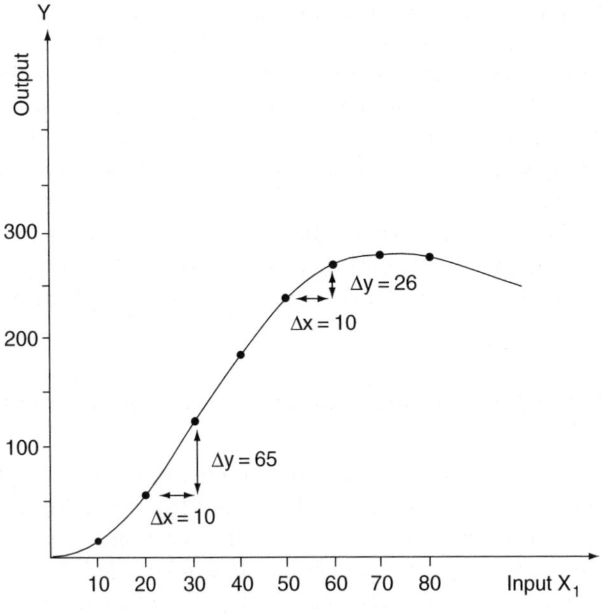

Figure 2.2 The total physical product curve

diminish, implying that additional units of inputs will eventually lead to small (that is, less than proportional) increases in output.

These notions are best summarized by using the concept of *marginal physical product* (MPP). The MPP for an input X_i refers to the change in the level of output associated with a change in the use of input X_i, *ceteris paribus*, that is assuming that all other remaining inputs remain fixed. It is expressed as follows:

$$MPP = \frac{\Delta Q}{\Delta X_i} \tag{2.5}$$

As units of the variable input are added to a set of fixed inputs, the MPP will eventually start to decline. Indeed, at lower levels of inputs (between 0 and 70), the MPP will be positive, implying for example that an extra pound of fertilizer per acre will have a positive impact on output, that is on cereal yields. For technical reasons, production will first increase at a rate above that of the use of fertilizer. However, as the farmer continues to add fertilizers to the production process, the increase in the yield becomes less and less important (after 30 units of input). That is, the MPP tends to fall. Eventually, intensifying the use of fertilizers beyond a certain level (after 80 units) will exhaust the soil, create pollution and impoverish the land, to such an extent that output will actually start to decline. This entails that the MPP will continue to fall until it becomes negative. This example illustrates the *law of diminishing returns*, also referred to as the *law of variable proportions*. The observation that agricultural output increases at a rate non-proportional to that of inputs was already present in the work of 18th century French economist Robert Turgot.

The *law of diminishing returns* has often been observed in practice, and it is central to the field of agricultural economics since it governs many of the conclusions concerning agricultural supply and resource use.

2.2.2 Production and supply functions – the objectives of the farm-firm

The farm-firm is a decision-making unit within the economy, and it generates a supply of agricultural products by employing a group of resources, factors of production or inputs. Combining the different inputs used by the farm, given a certain state of knowledge, will give rise to a certain level of output. The quantity supplied of an agricultural product is the amount of the product which will be offered for sale per time period under a given set of circumstances. The physical relationship between inputs and output is represented by the production function, which can, in its simplest form be written as follows:

$$Q = f(L, K, l) \tag{2.6}$$

where Q stands for the output produced within a specified time period, L is labor, K refers to capital and l is land. This production function describes the technological conditions of production in the farm-firm. The supply schedule of the firm, or the amount the firm actually plans to supply, is determined by a number of variables, including the technical relationship between inputs and the output, as represented by the production function. Therefore, we can write the farm-firm's supply function as follows:

$$Q = f(T, P_p, P_{1\ldots n}, I_{1\ldots m}, O, MS, W, G) \tag{2.7}$$

where:

Q is the quantity of an agricultural product supplied to a market per time period t,
T refers to the production function of the product as described above,
P_p is the price of the product,
$P_{1\ldots n}$ is the price of n other products, which could possibly be competing (such as in the case of substitutes),
$I_{1\ldots m}$ refers to the price of m inputs,
O represents the objective of the firm,
MS refers to the market structure relating to a specific agricultural industry. This refers in particular to the number of firms supplying the market, the size distribution of firms, and so on.
W is weather, and
G refers to government policy and to institutional factors.

The weather variable is obviously a factor that affects more specifically the output in the agricultural sector. It should be noted that variables G and P_p can be related. As will be seen in Chapter 3, government policy often takes the form of controlled prices, set normally at a level above market equilibrium. The theory of agricultural supply has often been based on the assumption that the farm-firm acts in such a way as to maximize profit. The objective of profit maximization is central to the neoclassical theory of the firm developed on the basis of the writings of British economist Adam Smith in the late 18th century, and formalized later in the work of Léon Walras (Walras, 1874). According to this view, the farm-firm will continue to expand output as long as the last (*marginal*) unit produced adds to total profit, that is as long as the revenue obtained from selling the last unit produced (*marginal revenue*) is greater than the cost of producing this last unit (*marginal cost*). Profit will be maximized when MR (marginal revenue) equals MC (marginal cost). With the development of alternative theories of the firm from the 1930s onwards, the neoclassical vision of the firm has weakened over time.[4] In particular, the following criticisms have been voiced:

(i) Marginal revenue and marginal cost are generally unknown concepts by firms, especially by small farm-firms. Moreover, it is argued that information is so imperfect and that uncertainty is such that farmers find it difficult to elaborate adequate production plans. However, as will be discussed in Chapter 5, farm management and farm planning agencies in Europe have tended to provide training to farmers so as to allow them to acquire the relevant knowledge with regard to this matter. Modern computer techniques in business and finance, such as the net present value technique, help businesses forecast more accurately future streams of profits. It is in turn argued that even if farm-firms have the knowledge of how to maximize profits, they are unsuccessful in doing so because of two non-predictable, hence non-measurable factors: weather and biological developments such as epidemics. Although it is true that modern forecast techniques help reduce the magnitude of uncertainty, the unpredictability of agricultural supply is still an important element for farmers around the world. For example, an econometric analysis of agricultural production trends in China has shown that weather as well as technological change emerged as the crucial factors shaping both short-run fluctuations and long-run stability of agricultural production in this country (Kueh, 1994).

(ii) Most farmers in the EU are not interested in profit maximization. Although it is acceptable to assume, for the purpose of theoretical simplification, that profit is the major and sole objective of the farm-firm, one must consider nevertheless all possible alternative objectives to the farm-firm. Other alternative objectives and motivations have been suggested, such as sales maximization and growth maximization. In particular, many small businesses are interested in growing, rather than in maximizing profit. Also, farming in the EU is often presented as being a lifestyle, rather than a pure business. Consequently, it has been suggested that the maximization of the farm household's utility may be a more appropriate objective for many farmers in the EU. Utility maximization is a broader concept than profit. In the farmer's utility function, there are various factors, apart from that of consumption of purchasable goods, that can be considered. These other factors are: leisure and farming as a lifestyle. Understandably, these factors may be difficult to measure, although the concept of opportunity cost can be used to that end.[5] The opportunity cost of a resource is the return it could earn in its best alternative use. Let us take for example the case of land, as a factor of production but also as a resource for the farmer. Assuming that the farmer owns his land, the opportunity cost of the land that the farmer works himself will simply be the forgone rent that the farmer could have received, had he decided not to farm it himself. It is however reasonable to assume that, for larger holdings, profit or sales maximization can be seen as the major objective of the firm. In addition, objectives may change over time. In a development economics perspective, the objective function of the typical Chinese farmer became household income maximization in the 1980s, whereas before that, it had been survival and self-subsistence.

2.3 The market mechanism for agricultural products

The market is a nexus of interactions between buyers and sellers. The exchange of products is the culminating point of the interactions between economic agents, and this exchange is made possible by the establishment of a price. The price variable is seen as a clearing mechanism, in that it permits to equate supply with demand. As seen above, the equilibrium price is set at a level that allows the last marginal unit supplied to the market to be sold. The objective of this section is to explain how supply and demand interact, so as to determine equilibrium prices and quantities of agricultural produce to supply. The supply–demand interaction can be:

- simple, that is from producer to consumer, or
- complex, that is from producer to final consumer, via food companies, who process, package, store, transport and distribute the product.

Markets are defined on a product as well as on a geographical basis. For any buyer, two products *x* and *y* belong to the same market, if they are *similar*, that is, if the purchase of *x* brings the same level of utility to the consumer as the purchase of *y*. Two products *x* and *y* are similar, if they are substitutes, that is, if their *cross elasticity of demand* is above 1. The use of various indicators enables the delineation of the geographical limits of a market (Jacobson and Andreosso-O'Callaghan, 1996). As a result, markets can be regional, national or global. Commodities such as cocoa, wheat, coffee, tea and vegetable oils are traded on a world scale, implying that their markets are global.

2.3.1 The market equilibrium paradigm

The economic world described by French economist Léon Walras is one where perfect competition prevails. Walras' general equilibrium model is used by economists as a reference paradigm, from which more complex real world situations can be understood. Although Walras has not brought anything new in terms of explaining economic phenomena, he was the first economist to propose a mathematical model of general equilibrium analysis. By using a system of equations, he shows that every economic agent, be it a firm or a simple individual selling its services (or labor) to a firm, obtains at equilibrium a maximum level of satisfaction. On each market the equilibrium is attained, and this leads to a general equilibrium situation which is the best possible outcome.

Five major assumptions underlie the model of pure and perfect competition. These are: atomicity, fluidity, homogeneity, rationality and free exit and entry.

(i) Atomicity means that there are many buyers and sellers on the market, implying that no one has the ability to influence the market price.
(ii) Fluidity suggests that producers and consumers have perfect knowledge of the conditions and events prevailing on the market.
(iii) Homogeneity entails that consumers are indifferent between the products of alternative suppliers. This means that there is no product differentiation in the market.
(iv) Economic agents are rational: firms are assumed to be profit maximizers, whereas consumers maximize their utility.
(v) Exit from and entry into the market is free, implying that firms face no barriers such as high capital requirements or sunk costs.

To illustrate this model, let us take a hypothetical demand and supply schedule as represented in Table 2.1 and Figure 2.3.

If, for some reason, a price is set arbitrarily at €2, producers will plan to supply 10 kg of strawberries per week, while buyers will be prepared to purchase 40 kg per week. At this proposed price level, a production

Table 2.1 Demand and supply of strawberries

Price (€/kg)	Demand (kg per week)	Supply (kg per week)
1	50	0
2	40	10
3	30	20
4	20	30

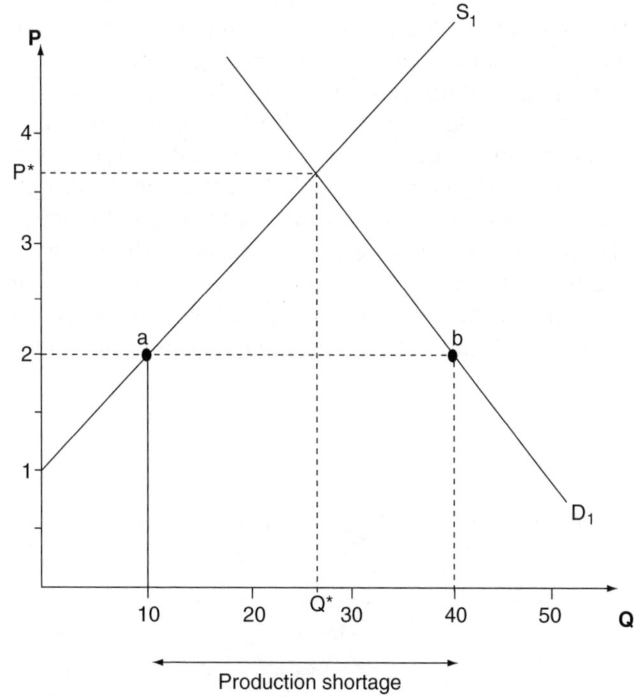

Figure 2.3 Demand and supply interaction on the market for strawberries

shortage will develop. Each producer will therefore find that he can increase the price, so as to satisfy consumers' demand. This will in turn enable the producer to increase his total revenue, and thus, his profit. As a result, the market price will be forced up, until it reaches its equilibrium level at P* (which according to the graph is around €3.5). At this price level, the equilibrium quantity is denoted by Q*, and this is the level at which supply equates demand. At this profit-maximizing level of output, the farm-firm covers its costs; in such a model, no supernormal profits are made. Conversely, if a price is set arbitrarily at €4 per kg, supply will exceed demand and stocks of unwanted strawberries will build up. As a consequence, the price will fall. This example shows how the price variable plays a clearing role in the market. If the price suddenly deviates from the equilibrium level of P*, a number of market forces will be set in motion until the equilibrium is restored.

If however the price is prevented from falling below €4 per kg, a quantity of 10 kg per week will remain unsold, and this will result in a long-term market surplus. In Chapters 3 and 4, we will see that there are several institutional, political and sociological factors that explain why some agricultural prices in the EU are *sticky*.

Although the two-dimensional graphical representation of price in relation to quantity, in line with the equation $P = f(Q)$, is convenient for the purpose of simplicity, we should bear in mind the fact that there are many other variables besides price affecting both demand and supply. Let us take the case of a change in one of the factors affecting supply, such as an increase in the price of a fertilizer input. Since less can be supplied at any price, the supply curve will shift towards the left from S_1 to S_2 (Figure 2.4). The new equilibrium price will be set at

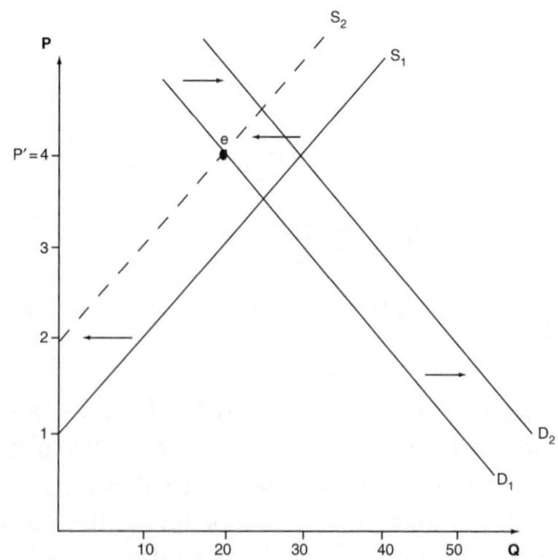

Figure 2.4 Changes in supply and demand conditions

€4 per kg of strawberries. As before, this change in supply conditions takes place *ceteris paribus*, that is by assuming that no other variable other than the price of fertilizer changes.

Similarly, changes in the factors affecting demand, other than the price of the product, can cause the demand curve to shift either way. Assume that consumers' tastes change in such a way as to shift the demand curve towards the right (from D_1 to D_2 in Figure 2.4). The change in consumer habits and tastes has been an important feature of economic integration in the EU. The fears of oversupply in several agricultural markets have been atrophied by a formidable change in consumer's diet. This is particularly true for olive oil, a product that many north-European consumers have discovered only in the past thirty years or so. The change in the demand conditions implies that more will be demanded at the same price. This could develop into potential shortages. Two possible reactions could ensue: (i) in the very short run, the price will continue to assume its clearing function, and a new equilibrium will be reached, with a higher price P′ (determined by the intersection between D_2 and S_1); (ii) the potential short-term shortage and consequent price increase (to P′) is taken by the firms as a signal that profits can be made in the strawberry business. Higher prices are indeed an incentive for potential entrants to enter the market, and/or for incumbent firms to raise their output levels. Either way, these reactions will shift the supply curve to the right, leading to a decrease in price (not shown in the graph). This second scenario shows that in the long run, supply adapts to evolving demand patterns. This implies that supply is perfectly elastic in the long run.

2.3.2 Short-term price fluctuations – the cobweb theorem

Many markets for agricultural products are notoriously unstable. In the short run, food commodity prices vary widely and erratically. In the long run, producer food prices in the EU and in the world have trended down for several years and decades.

In the short run, the erratic variation of prices is explained by two supply characteristics prevailing on agricultural markets. The first characteristic relates to the fact that the production of many agricultural products is subject to natural and uncontrollable factors which cause unplanned variations in output. In spite of many advances made in meteorology, biology and technology in general, the weather is still very much a random phenomenon, and the outbreak of many plant and animal diseases such as the recent BSE (bovine spongiform encephalopathy) and foot-and-mouth disease is still not yet under human control. As noted earlier, the supply curve shows the quantity that producers *plan* to supply to a market at alternative prices. It follows that, because of these two supply elements, for example, weather and diseases, the actual quantity supplied typically and inevitably differs from the planned quantity supplied. In addition, prices will play their clearing function in equating the quantity demanded with the actual quantity supplied, albeit this will happen only up to a certain extent. If for example a market is characterized by an exceptionally good year, prices will decrease, but demand will not increase accordingly (or proportionately), because of the price and income inelasticity characterizing many agricultural markets in the EU. The second supply characteristic is related to the production time lag existing in agriculture. A typical agricultural market is one in which there is a time lag, of sufficient length to encompass the biological life cycle of plants and animals, between the decision to produce a certain level of output, in t_0, and the produce becoming available for supply on the market, in t_1. The decision to produce a certain level of output in time period t_0 is in turn influenced by the price prevailing in t_0. Therefore, we need to refer to a model of an agricultural market, where supply curves describe the relationship between output in the current time period and price in the previous time period. This model is known as the *cobweb model* and it is based on the *cobweb theorem*. The cobweb theorem relates to a market for which the conditions of pure and perfect competition apply, supplemented with the assumption that producers respond to the previous period's price, given the time lag between planned and actual output. The phenomenon of lagged output curves, as the model is sometimes referred to, was first analyzed by German economist Hanau in the case of the pig market. Figure 2.5 shows how an initial plan to produce Q_0 at time period t_0, based on a price P_0, can lead to a series of reactions and adjustments, until the equilibrium (P*, Q*) is reached.

Figure 2.5 represents the specific case of a converging pattern. This is explained by the fact that the slope of the supply curve is steeper, or more inelastic, than the demand curve. If we had a situation where the slope of the demand curve is steeper, or more inelastic, than the supply curve, the result would be a diverging pattern, with the inability of the system to converge towards equilibrium. Consequently, the less price elastic the demand, with respect to supply, the more likely the diverging pattern. As we have seen before, this is likely to happen in the EU as consumers have been found to be not much responsive to price changes.

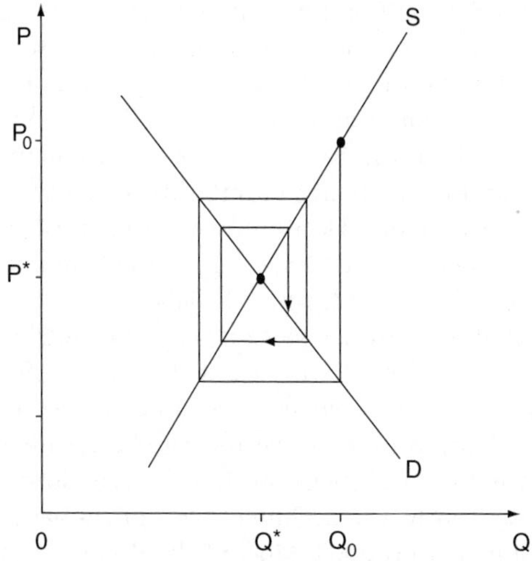

Figure 2.5 Disequilibria on agricultural markets – the cobweb pattern

2.3.3 Long-term price fluctuations

The 'secular' decline of agricultural prices at the world level – when compared with manufacturing prices – has been documented by many authors (see for example Spraos, 1980; Sapsford, 1985; Grilli and Yang, 1988; Anderson, 1990). At the world level, an examination into the movement of the terms of trade for agricultural products over the long term shows a high variation of agricultural prices, relative to prices of other products. The long-term trend is that of a gradual decline, slightly offset by rapid short-term rises. There were for example historical peaks in the mid-1970s. Since that time, agricultural prices have declined continuously, with cereal prices, and in particular rice prices, falling the most.

At the EU level, the dominant long-term trend has been again a substantial fall in real farm prices almost across the board. When compared with general prices, prices received by farmers in the Union's member states fell substantially between 1950 and the 1980s, that is between the time of the gestation of the Common Market and that of the beginning of the multilateral trade negotiations under the Uruguay Round (see Chapter 8). Obviously, variations from period to period can be observed. Between 1960 and 1965, farm prices in the founding member states fell substantially less than during the transition period

(1965–70). In the first half of the 1970s, real prices actually rose in Italy, the UK and Ireland. For the latter countries, this was explained by a realignment of domestic prices towards the Community level. During the 1970s, The Netherlands and Germany were the two countries where producer prices fell most. Favorable price movements in countries such as France and Luxembourg during this period were partly due to exchange rate movements. Declining producer prices continued well into the last decades. In the 1980s, agricultural production increased by an annual rate of 2 per cent, while the labor force engaged in the agricultural sector contracted by 2.5 per cent per year (EUROSTAT, 1991). Food prices, that is those prices determined at the end of the producer–final consumer chain, have also trended downwards in real terms. However, the food price to farm price ratio has increased generally in all countries of the EU since the late 1960s, with the exception of Ireland and Denmark. Consequently, the decline in farm prices has only marginally benefited consumers in the EU. Most of the decline in producer prices has not been passed on to final consumers in the EU, or put in another way, this decline has greatly helped the EU food industry increase its profit margins.

The factors responsible for this long-term decline in the EU, and elsewhere in the world, are an inelastic demand, price and income wise, and a substantial expansion of agricultural production in spite of a declining farm labor force. This expansion has been made possible by technical progress. Inventions and innovations in the chemicals, machinery and transport industries as well as in the power generation sector have permitted many cost minimizing new techniques, as we will discuss in Chapter 5. Gains in terms of labor productivity have been the most significant feature of EU agricultural markets since the 1960s.

Have these productivity gains been translated into higher incomes for the farming population? If we take an index measuring the agricultural terms of trade,[6] defined as the ratio between real output prices and real input prices, we notice that this index has fallen consistently over a long-term period. For example, this ratio declined from a level of 100 in 1990 for the Union as whole to 91.1 in 1996 (EUROSTAT, 1996). It follows that the profit of EU farmers, as expressed by the simple equation $\pi = TR - TC$, has generally been squeezed by relative increasing input costs. Although in theory lower farm and food prices benefit consumers – an important consideration for developing economies and also for the less well off consumers in EU countries – they also depress farmers' incomes. Net incomes from agricultural activity in the EU have generally trended downwards (Hill, 1996). The long-term relative decline of farm prices and incomes has nurtured the so-called 'farm problem'.

2.3.4 Do the agricultural markets conform to this idealistic paradigm of perfect competition?

The existence of negative profits and the fact that there is no normal return on the factors of production make agricultural production a specific economic activity in that it does not operate on the basis of profit maximization. In many cases, agricultural production is organized on the basis of the family firm, where the objective is simply to optimize the value of the labor force employed (EUROSTAT, 1991). Moreover, the discussion above has highlighted the difficulty to reach an equilibrium in the case of many agricultural products, given the specific supply and demand conditions prevailing on agricultural markets. It is worth therefore analyzing whether the perfectly competitive framework applies to the general field of agriculture. Undoubtedly, many traditional agricultural products are typified by the interaction of many buyers and sellers, by extremely little product differentiation, as well as by insignificant barriers for new potential entrants. This is the case, for example, for the production of potatoes, carrots, milk and fruits. Entry barriers in these markets are unquestionably much lower than those existing in say, the shipbuilding, aerospace or even computer industries. In addition, although there is no perfect knowledge on many agricultural markets, many farmers are well informed about prices prevailing in alternative markets. Consequently, it may be implied that more agricultural markets approximate the perfectly competitive market paradigm than do markets for manufacturing goods. However, developments over the years have meant that the reference to this simplified form of market structure is however less and less adequate. In the last decades, EU agriculture has been witness to the emergence of large farm holdings and institutions providing marketing services. In more recent years, issues in terms of consumer safety have become prominent and have increased the degree of product differentiation in agricultural markets. The increased popularity of organic farming has meant that even in traditional markets such as potatoes and carrots, production differentiation is an important strategy for farm-firms and that the perfectly competitive model applies less and less.

It is therefore necessary to envisage the existence of alternative market structures as suitable frameworks for the study of modern agricultural markets in the EU, an issue to which we now turn.

2.4 The structure of EU agricultural holdings

Market structure refers to the intensity of competition on a given market; it embraces many different situations lying between the two extremes, for example perfect competition, as seen above, and monopoly. Monopoly, oligopoly and monopolistic competition refer to situations of 'imperfect competition'. The theory of imperfect competition was developed simultaneously on both sides of the Atlantic by Chamberlin (1933) and Robinson (1933). During the 1920s, the work of American economist Clark (1923) and that of Italian economist Sraffa (1926) had laid down the basis for these theoretical developments. Market imperfections arise essentially from agglomeration effects, from product differentiation, as well as from the uneven distribution of output amongst firms in the market. The unequal distribution of output is explained by the coexistence of small and large producing units in the same industry. In 1998, the average size of agricultural holdings in the EU ranged from 6.1 ha in Greece to 130.9 ha in the UK (Table 2.2).

Table 2.2 Size of agricultural holdings in the EU (1997/98)

	Average size of holdings (ha)	**Percentage of small agricultural holdings***
Belgium	33.3	–
Denmark	48.9	–
Germany	52.7	–
Greece	6.1	62.7
Spain	29.7	49.0
France	62.1	–
Ireland	37.1	41.4
Italy	12.3	54.4
Luxembourg	58.6	–
The Nertherlands	25.8	–
Austria	24.1	–
Portugal	12.2	16.7
Finland	32.8	–
Sweden	83.4	–
United Kingdom	130.9	1.3
EU-15	30.9	38.0

* A 'small' agricultural holding is one with eight or less ESU (European Size Unit); this is a harmonized European measure of size. By contrast, a very large holding is one with 100 or more ESU.

Source: The Agricultural Situation in the European Union (2002).

Table 2.3 Concentration in selected EU agricultural industries (1999)

	Cattle farms	Pig farms	Dairy cows
Very large farms			
% of total animals	52.2	51.8	16.5
% of farms	13.3	2.6	2.3

Source: The Agricultural Situation in the European Union (2002).

The third column of the table shows that the smaller agricultural holdings in the EU represented still 38 per cent of all EU agricultural holdings in 1997/98, with variations from 62.7 per cent in Greece to 1.3 per cent in the UK. Small size, as defined by the agricultural area, commands both the level of output and productivity. For example, in the cattle industry, more than half of all EU animals were concentrated among only 13.3 per cent of farms in 1999. These were very large units with more than 100 heads of cattle. The corresponding figures in the dairy cow sector were 16.5 and 2.3 per cent respectively. The level of concentration is even higher in the pig sector where only 2.6 per cent of farmers produced more than half of all animals in the EU (Table 2.3). These very large farm units typically have more than 1000 animals. Owing to economies of scale and to other factors such as higher levels of innovation, large agricultural holdings in the EU tend to be more productive. In 1997/98, the average value added per holding was 7.5 times greater for the group of very large holdings, compared with the small ones (CEC, 2002).

2.5 Concluding remarks – the 'farm' problem

Since the 1960s, agricultural markets in the EU have been characterized by a declining, and at best, a static demand, as well as by oversupply. On the demand side, a low and declining income elasticity of demand for food, combined with a low price elasticity of demand, has placed a limit to the expansion of demand. On the supply side, technical progress in the power generation, chemicals and long-distance transport sectors has permitted cost minimizing new techniques, leading to a substantial expansion of agricultural output, in spite of a declining farm labor force. Technical change has also resulted in unprecedented increases in total factor productivity in EU agriculture. Although most of post-war output growth in EU agriculture is explained by technological change (see Chapter 5), the *law of diminishing returns* warns against the limits set by input intensification. The interplay of demand and supply characteristics explain why producer prices and incomes have declined over the last decades; this situation epitomizes what is known as the 'farm' problem.

Agricultural producer prices have declined relative to industrial prices during the 20th century. In the EU, most of this decline has not been passed on to the final consumer. In the general context of post-war rapid economic recovery and growth, farm prices, and by extension farm incomes have failed to keep pace with other incomes. These arguments concur to support the view that there has been a decline in the terms of trade in agriculture.

This chapter has also highlighted the specificity of agricultural markets, in that, when compared with most manufacturing markets, forecasting supply with any degree of accuracy is made difficult by the production time lag inherent to the production of most food products. Agricultural markets are consequently notorious for being unstable, and for diverging substantially from the market equilibrium paradigm formalized in the late 19th century by Leon Walras. Modern EU agricultural markets are characterized by the coexistence of small and large production units, by product differentiation strategies increasingly reinforced by concerns over food safety.

Key terms and concepts

Demand for agricultural (food) products
Supply of agricultural products
Price elasticity of demand
Income elasticity of demand
Engel's law
Law of diminishing returns
Market equilibrium
The 'farm' problem
Size of agricultural holdings and concentration

Notes

1 There are exceptions to the general rule of an inverse relationship between the quantity demanded and the price. Giffen goods, for which the income effect exceeds the substitution effect, are characterized by an upward sloping demand curve. In some cases, an increasing price may be taken as an indi-

cator of superior quality. Finally, in times of high inflation, a rise in price might be taken by consumers as a signal that further price increases are likely to follow.

2 Product differentiation refers to the consumer's ability to distinguish between competing products. From the viewpoint of the firm, this entails enhancing the non-price attributes of products through sales efforts.

3 A distinction needs to be made between income elasticity of quantity demanded and income elasticity of expenditure. Expenditure includes the quality aspect, meaning that as income rises, the consumer will switch to higher quality and higher priced varieties. In this case, income elasticity of expenditure will exceed income elasticity of quantity demanded.

4 For more on the criticisms of neoclassical thinking and on the alternative theories of the firm, see Jacobson and Andreosso-O'Callaghan, 1996, Chapter 3.

5 The concept of opportunity cost will be developed more fully in Chapter 7.

6 The Terms of Trade refer to the rate at which units of one product can be exchanged for units of another product.

Bibliography

Anderson, K. (1990) *Changing Comparative Advantages in China – Effects on food, feed and fibre markets*. Paris: OECD Development Center.

CEC (2002) *The Agricultural Situation in the European Union – 2000 Report* (Brussels and Luxembourg).

Chamberlin, E.H. (1933) *The Theory of Monopolistic Competition* (Cambridge, MA: Harvard University Press).

Clark, J.M. (1923) 'The economics of overhead costs', *The Journal of Political Economy*, 31: 5, October, 606–36.

EUROSTAT (1991) *Generation and Distribution of Productivity Increases in European Agriculture*, 1967–87 (Brussels and Luxembourg).

EUROSTAT (1996) *Agricultural Prices* (series 5B, 1996/4) (Brussels and Luxembourg).

EUROSTAT (2002) *Yearbook – The Statistical Guide to Europe, Data 1999–2000* (Brussels and Luxembourg).

Grilli, E.R. and Yang, M.C. (1988) 'Primary commodity prices, and terms of trade of developing countries: what the long run shows', *World Bank Economic Review*, 2: 1, January, 1–48.

Hill, B. (1996) *Farm Incomes, Wealth and Agricultural Policy* (Sydney: Avebury, Aldershot).

Jacobson, D. and Andreosso-O'Callaghan, B. (1996) *Industrial Economics and Organization – A European Perspective* (London: McGraw-Hill).

Johnston, D.G. (1991) *World Agriculture in Disarray* (London: Macmillan).

Kueh, Y.Y. (1994) *Agricultural Instability in China, 1931–1990: Weather, Technology and Institutions* (Oxford: Oxford University Press).

Robinson, J. (1933) *The Economics of Imperfect Competition* (London: Macmillan).

Sapsford, D. (1985) 'The statistical debate on the net barter terms of trade between primary commodities and manufactures: a comment and some additional evidence', *Economic Journal*, 95: 379, September, 781–8.

Spraos, J. (1980) 'The statistical debate on the net barter terms of trade between primary commodities and manufactures', *Economic Journal*, 90: 357, March, 107–28.

Sraffa, P. (1926) 'The laws of returns under competitive conditions', *The Economic Journal*, 36: 144, December, 535–50.

Walras, L. (1874) *Eléments d'Economie Politique Pure ou Théorie de la Richesse Sociale* (reprinted 1952; Paris: Librairie Générale de Droit et de Jurisprudence).

3
Government Intervention in Agriculture

Objectives of this chapter

This chapter is intended at:

- Presenting the economic rationale for government intervention in agriculture.
- Analyzing the main economic policies affecting the agricultural sector.
- Presenting a brief historical review of government involvement in agriculture in a selected number of EU countries prior to the implementation of the Common Agricultural Policy (CAP).

Introduction

Food and agricultural policies are aimed at influencing the behavior of the agricultural sector and that of agricultural markets therein. Any economic policy consists of two major elements: the *objectives*, based on a given rationale, and the *instruments*, which are devised to achieve the set objectives. With regard to agricultural policies, different countries in the world have different objectives and they emphasize different constituent elements of the various policies. In less developed countries, the emphasis has normally been placed upon stimulating agricultural exports so as to earn foreign exchange, and on securing indigenous inputs, such as cotton, for the local development of agricultural processing and manufacturing industries. Agricultural policies are of two broad categories: the first category relates to agricultural-support policies that involve normally the manipulation of agricultural prices. Subsidies to producers, as in the case of the EU, or to consumers, as in the case of developing countries, are normally devised on a year-to-year basis. The second category entails devising structural policies. Spanning normally over a long period of time, these policies are aimed at affecting the structure of the agricultural holdings, by rendering them more efficient. One way in which structural policies have radically changed the nature of the agricultural sector in some European countries, is through modernization in Southern Europe, and through privatization in the former centrally planned economies of eastern and central Europe (see Chapter 6). Structural policies encompass investment grants and public investment in education, research and infrastructure; they can also take the form of retirement schemes and of land reform measures.

3.1 The rationale for government intervention in agriculture

Most countries in the world have in all times supported their agricultural sector. One of the earliest written accounts of government intervention in agriculture is contained in the Ancient Testament of the Bible[1] (Genesis, 41:1). According to the scriptures, the Pharaoh's central administration policy of withholding one-fifth of each producer's food supply during years of abundance, and in order to ensure regular agricultural supplies, became law in ancient Egypt. In modern developed countries, such as the EU and the USA, government intervention has been aimed at alleviating the social strains brought about by economic change. The objectives of agricultural policies in these countries have been to stabilize farmer's incomes and also to slow down the migration out of the sector. In the developing world, intervention helps lessen the perceived (and real) instability of international markets for agricultural commodities. For example, rice has traditionally represented a high percentage of export earnings in countries such as Thailand. Since rice is also the main staple for consumption, and since the rice market is very unstable at the world level, the Thai government has tried to insulate the domestic market from world fluctuations, through the taxation of rice exports (Colman and Young, 1993). In developing countries, importables are normally protected (Krueger *et al.*, 1988).

Moreover, and as we have seen in Chapter 2, supply and demand characteristics prevailing in agricultural markets make the agricultural sector different from other industries or sectors in any economic system. The high unpredictability of supply, due to weather conditions and to biological developments, makes intervention desirable. In the case of the EU, economic considerations explain nevertheless only partly why governments intervene on behalf of agriculture. There are non-economic arguments that reinforce the 'specific' nature of agriculture; indeed, socio-political as well as strategic considerations also play an important role in shaping agricultural policies. From a socio-political point of view, farmers in Europe used to form a cohesive and large social class that conveyed traditional moral values. When the Common Market was formed in the second half of the 1950s, farmers represented 39 and 32 per cent of total employment in Italy and France respectively. This relatively large and conservative class was instrumental in keeping the communist threat away from post-war Western Europe. For example, the 'vote paysan' in France was an important contributory force in the re-election of several right-wing governments and Presidents until the Presidential election of 1981. Consequently, the protection of farmers' incomes from the effects of world competition, and from their inexorable downward trend, as explained in the previous chapter, has been an essential objective of farm policies throughout Europe. From a strategic angle, food shortages during times of conflict in Europe have made it clear that agricultural production should be maintained in anticipation of a developing world food shortage, so as to decrease the reliance on imported foodstuffs. Even the country that gave birth to the industrial revolution, that is the UK, has been heavily concerned with the strategic advantages of maintaining a relatively large domestic agricultural sector. Before we look at the way various European countries have dealt with the protection of their farming sector, we will first deal with a number of agricultural support policies.

3.2 Agricultural support policies: methods and effects

This section deals exclusively with agricultural support policies. These policies encompass price support and production support, although price support methods have been the most common practice. Price support policies entail price stabilization. This can be done by lowering the agricultural prices when they are high and by increasing them when they are low. It can also imply raising the prices when they are judged low and leaving them unchanged when high. The aim of price stabilization policies is again to prevent an undesirable fluctuation in farmers' incomes. These price stabilization methods are normally accompanied with an increase in the general price level in the domestic economy. A possible classification of instruments affecting the agricultural sector is proposed in Table 3.1. The table lists the various possible instruments according to whether they are imposed: (i) directly at the farm level, (ii) at some point in the domestic market, (iii) at the frontier

Table 3.1 Classification of selected policy instruments

Farm	**Domestic market**	**Frontier**
Deficiency payments	Public intervention: buying up excess of supply; building up stocks	Import tariffs and variable levies
Production subsidies	Food subsidies to consumers	Export subsidies
Input subsidies/credit	Indirect taxation	Import quotas
Production or acreage quotas	Grants and loans to farming industry	Non-tariff barriers
	Public investment in research	

Source: Derived from Colman and Young (1993: 269).

with other economies. For the purpose of any EU country, the 'domestic market' is equivalent to the EU market. The frontier considered here will therefore be the one separating the EU agricultural market from that of the rest of the world.

The instruments applied at all three levels distort production and price patterns from levels generated by competitive pressures in national and international markets. The *frontier-level* instruments alter the volume of trade flows, by normally restricting the amount of imports. On international markets, the distortion may affect the output and price levels of other supplying countries, depending on the size of the protected economy. In this section, we will devote a special attention to the deficiency payments system, since it has been a system favored by some European economies before they joined the EU, as well as by the USA. The latter part of the section will highlight protection at the edge of the domestic market; we will look at import quotas, tariff protection and variable levies.

3.2.1 Selected policy instruments

3.2.1.1 The deficiency payments system

A deficiency payment is a variable subsidy paid per unit of output to compensate for the shortfall (or deficiency) between the average equilibrium price and a higher guaranteed price, which is judged desirable by the government. The level of the guaranteed price is set to a level compatible with the government's objective of price or income stabilization. Figure 3.1 depicts the supply and demand conditions prevailing in one specific agricultural market of a home country *H*. At price P_1, domestic demand is OQ_1 and domestic supply is OQ_2 generating

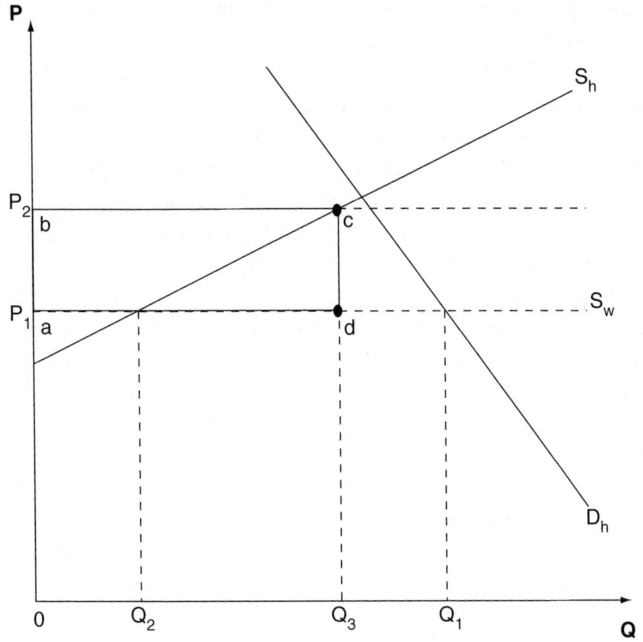

Figure 3.1 The deficiency payments system

imports equal to $OQ_1 - OQ_2$. The introduction of a subsidy per unit of output will boost domestic output and consequently farmers' incomes. A subsidy of $P_2 - P_1$ will raise domestic supply to OQ_3, and will restrict imports to $OQ_1 - OQ_3$. The consumer is not affected by the subsidy in that he still enjoys the low price, whereas the taxpayer will bear the burden of the policy (area *abcd*). The farmers' income is made up of a market-determined portion, equal to area OQ_3da, and of a subsidized portion (area *abcd*).

With this policy, the government's expenditure is unpredictable, given that the unit deficiency payment depends also on world-market prices.

3.2.1.2 *Import quotas*

Import quotas and VERs (voluntary export restraint agreements) are part of a broader set of non-tariff barriers (NTBs). These quantitative restrictions are protective instruments that can be applied at the frontier, so as to set a limit on the quantity (or value) of a product permitted to enter or leave a country. In Figure 3.2, the domestic supply curve of a commodity is represented by S_h whereas S_m represents the domestic demand for foreign imports of the same commodity. Summing S_h and S_m gives the total supply curve S_{h+m}. In the absence of discrimination against imports, and assuming that the consumers view the domestic and foreign products as perfect substitutes, the equilibrium price and quantity will be set at levels P_1 and OQ_1. At that price, domestic supply will be OQ_{h1} and imports will cover the excess of demand over supply. These imports are represented by M_1. If the domestic government decides to protect the national industry by imposing an import quota, which is equivalent to a predetermined quantity, the supply curve for foreign products becomes the kinked curve denoted by S'_m. The total supply curve becomes therefore S'_{h+m}. It can be seen that the intersection between the new total supply curve and the demand curve D helps determine a new equilibrium. The new market price rises to P_2, making consumers worse off. At that price, total demand declines to OQ_2.

Compared with the previous situation, which was one of free trade, domestic supply increases, whereas foreign supply (or imports) contracts.

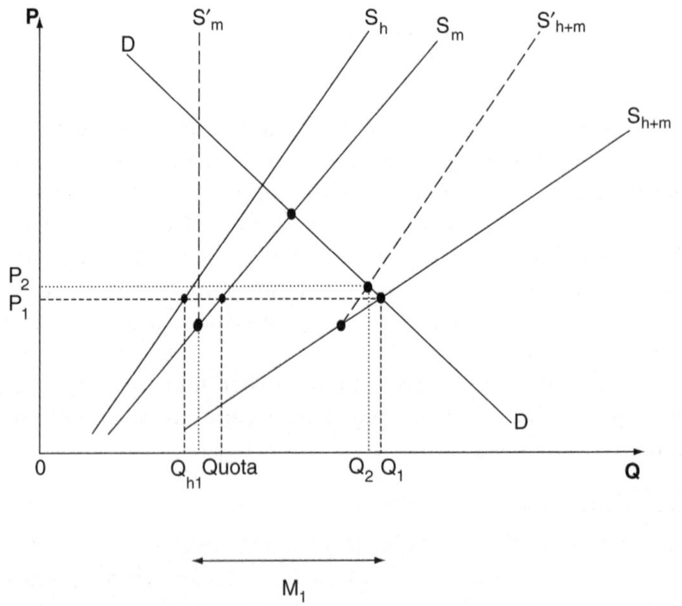

Figure 3.2 Import quotas

3.2.1.3 *Tariffs*

A tariff is equivalent to a tax on importing a good or service into a country, and is collected by customs officials at the place of entry. Tariffs can be specific or *ad valorem*. A specific tariff relates to a monetary amount paid per physical unit of import, whereas an *ad valorem* tariff is calculated as a percentage of the estimated market value of the goods when they reach the importing country. Figure 3.3 depicts the situation of a government in country *H*, discriminating against foreign produce by imposing a tariff on imports. The tariff shown in this diagram is the case of a constant amount paid per physical unit of import. In a situation of free trade, that is at the price level P_w, a quantity $OQ_1 - OQ_2$ is imported (denoted by M_1). P_w is the price prevailing on the world market; this corresponds to the price level of the most competitive producing unit in the world. Any farm-firm with a cost structure, and

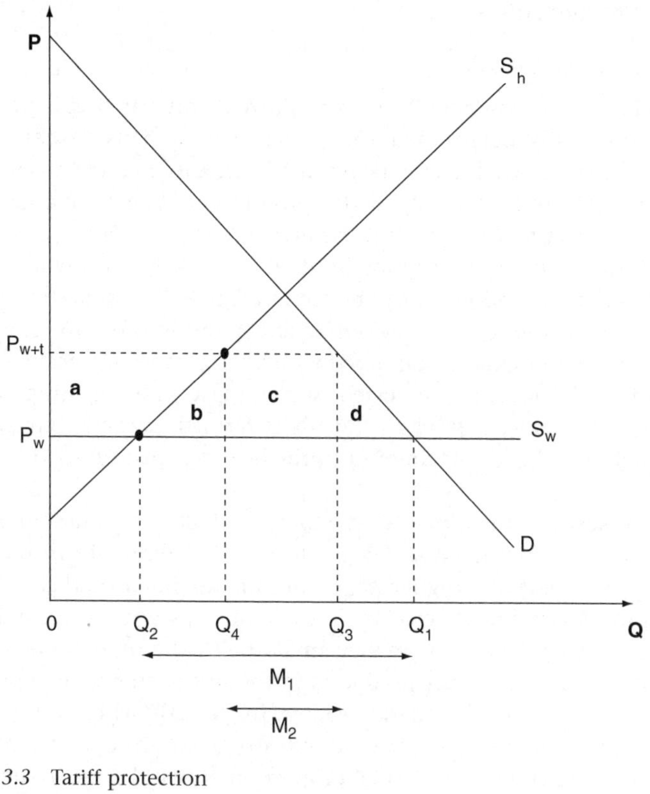

Figure 3.3 Tariff protection

therefore a price, above P_w is economically non-viable. At price level P_w, the world supply curve S_w is horizontal or perfectly elastic. The assumption of perfect elasticity of the world supply curve implies that this price is not affected by changes in the supply or demand schedules of country *H*. The imposition of a tariff raises the price to P_{w+t}, and leads to a decrease in the quantity demanded (OQ_3).

A higher market price is again interpreted by domestic producers as a signal that profit opportunities exist in the industry concerned, and this will lead to an expansion of the domestic supply from OQ_2 to OQ_4. At this new equilibrium level, imports have clearly contracted; a quantity M_2 is now being imported. Consequently, the tariff represents an opportunity for domestic suppliers, and it penalizes foreign imports by taxing them. Faced with an increase in the price of foreign and competing imports, domestic consumers turn increasingly to domestically produced commodities. Domestic producers enjoy consequently the benefits of extra sales and of higher prices, on competing imports, thanks to the tariff.

3.2.1.4 *Variable levies*

The objective of variable levies is to prevent imports occurring below some politically determined and desirable level. This desirable price level is often referred to as a 'target' price (see Chapter 4). In Figure 3.4, the price P_2 is indeed judged by the government as being in accordance with its objective of stabilizing farmers' incomes. A target price of P_2 will stimulate domestic supply, from OQ_{h1} to OQ_{h2}, and will penalize domestic consumption. As can be seen in Figure 3.4, domestic demand will contract from Q_1 to Q_2. Consequently, and in order to respond to the decrease in demand combined with an expanded domestic supply, imports will have to be levied accordingly. The resulting import levy will be set at a level $P_2 - P_3$. It is termed 'variable' given that, although P_2 is fixed, the amount of the levy depends on world market conditions.

At the level of the domestic economy, variable levies raise domestic prices above international levels. This instrument allows domestic producers to enjoy a higher price and an expanded market, whereas world producers are affected by a lower price P_3 and by a contracted market. The costs of such a policy are borne by the final domestic consumer, as well as by other producers in the rest of the world who have to face lower prices and market shares. In the case where the country protecting its domestic agricultural industry by means of a variable levy system accounts for a substantial proportion of world trade in the given

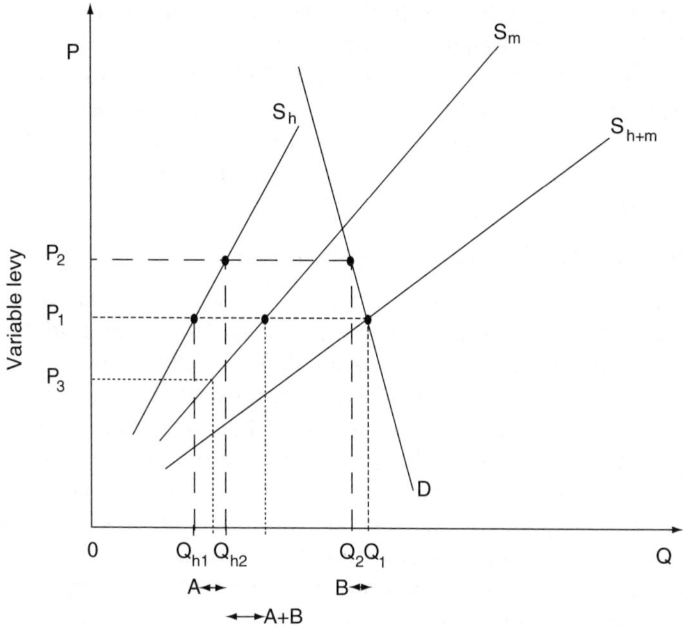

Figure 3.4 Variable levies

commodity, then this policy will tend to depress world prices. In Chapter 4, we will see that variable levy systems are generally used concomitantly with market intervention policies.

3.2.1.5 Other non-tariff barriers (NTBs)

NTBs are a modern form of trade impediments. NTBs are of different kinds, and encompass broadly a set of quantitative restrictions, import licensing schemes, antidumping and countervailing duties, quotas, VERs, input and food subsidies, as well as institutional or legal measures such as phytosanitary and veterinary regulations. The phytosanitary regulations are all government regulations that restrict or prohibit the importation of certain plant species, or products of these plants, so as to prevent the introduction of pathogens that these plants may be carrying. In the same way, sanitary regulations restrict or prohibit the importation of certain animal species, or products of these animals, so as to prevent the introduction of pests or diseases that these animals may be carrying (OECD, 1995). Some geographically remote countries have argued that these measures are essential to guarantee their disease-free status. This has been the case of the Republic of Ireland in its preparation to the Single European Market. In many cases, these NTBs have however been used as a subtle way to hamper trade in sensitive sectors such as beef and dairy products. In recent years, the development of genetically modified organisms (GMOs) has tended to increase awareness *vis-à-vis* NTBs (see Chapter 5).

3.2.2 The effects of the policy instruments

The effects of protection in the agricultural sector are manifold. The different policy instruments affect prices, consumption and production patterns (domestically and internationally), as well as national budgets and balance of payments. In addition, as most of the policies tend to favor domestic production, the result has often been an intensification of production, leading to the well-known detrimental effects on the environment. Table 3.2 summarizes synoptically the essential impact of a few selected policies, on consumption, government's expenditure, prices and production (domestic and international).

As can be seen, all policies stimulate domestic production by penalizing imports from the rest of the world, as well as either domestic consumers (as in the case of import quotas and variable levies) or domestic taxpayers (as in that of deficiency payments). An important impact of these policies is also the one on world prices, an essential concern in an era of world trade liberalization. Variable levies imposed

Table 3.2 The effects of agricultural protection

	Deficiency payments	**Import quotas**	**Tariffs**	**Variable levies**
Domestic production	Increases	Increases	Increases	Increases
Imports	Decreases	Decreases	Decreases	Decreases
Domestic consumer prices	Decreases	Increases	Increases	Increases
World prices	Lower	Higher	–	Lower
Domestic production prices	–	Higher	–	Higher
Domestic consumption	Increases	Decreases	Decreases	Decreases
Budgetary expenditures	Increases	–	Extra revenue	Extra revenue

by a large agricultural producer and trader, such as the EU, would tend to depress world prices. As we will see in the next chapter, this is a system that has been favored by the EU for many years.

Let's look at the case of a tariff. By taxing imports, a tariff generates revenue to the government. This revenue equals the unit amount of the tariff, multiplied by the volume of imports subject to the tariff in question. This is represented by area *c* in Figure 3.4. The tariff represents a gain for the nation as a whole. This gain can eventually be redistributed to the farming population, through investment subsidies, so as to stimulate the modernization of agricultural holdings. It can also be matched by an equal cut in some other tax, such as VAT, so as to lessen the negative effect on consumers. It should however be noted that part of this extra revenue will probably be lost through the administration of the tariff itself. In addition, by taxing foreign imports, the tariff brings gains to domestic producers who face import competition. This gain is represented by area *a* in Figure 3.4. Areas *b* and *a* are known as representing the 'production' and 'consumption' effect respectively. On the producer's side, a protected market leads to a shift from cheap imports to more expensive domestic production. Indeed, as seen above, the tariff raises domestic production, and this is done at the expense of imports. A price P_{w+t} which is well above the price prevailing in a free trade situation (that is P_w), entails that production patterns in the home country *H* tend to become inefficient. In other words, and compared with a situation of free trade, additional production factors in *H* are needed so as to produce the extra output denoted by $OQ_4 - OQ_2$. These factors are drawn away from other uses and they are now utilized in this protected, and increasingly inefficient, industry. Had they been employed elsewhere, these factors of production would have led to an optimal specialization pattern for *H*. The tariff generates therefore a shift in the production structure of country *H*, involving a cost that exceeds the savings made in not importing an extra amount of the product concerned. The shift to more expensive home production is called the 'production effect' of the tariff; it is represented by area *b*. In addition, consumers lose out, as a result of this protective policy. Their net (or deadweight) welfare loss is calculated as the decrease in the quantity consumed $OQ_1 - OQ_3$ times the gradually increasing price. This is represented by area *d*. This 'consumption effect' shows the loss to consumers in the importing nation, who now have to cut their consumption. Finally, this simple diagram shows also that there is a redistribution of wealth from consumers to producers (represented by area *a*). The loss of revenue suffered by the exporting country (the rest of the world) is not represented in this model. To summarize, the effects generated by the imposition of a tariff are: a gain for the government (area *c*), a gain for producers (area *a*), a loss for producers (area *b*), and a loss for consumers (areas $a + c + d$). Consequently, the *net* effect of a tariff is a loss for the nation as a whole, and this is represented by areas *b* and *d*. These are termed 'deadweight losses', implying that neither the government, nor producers nor consumers gain it.[2]

3.3 Early European agricultural policies

Contemporary EU agricultural policy is embodied in the CAP (Common Agricultural Policy). The CAP was born in the late 1950s and has become widely notorious for creating many distortions at both the EU and international levels. However, the CAP did not create government intervention in the agricultural economies of Europe. In most of the EU countries, agricultural support had been in existence well before the CAP came into effect. The CAP merely transferred arrangements existing at national levels, into a more harmonized European framework. A brief historical presentation of government involvement in the agricultural sectors of the EU countries will help understand the context in which the CAP was born, as well as the main features it embodied.

3.3.1 Developments before the First World War (1914–18)

Although agricultural trade was confined to a small proportion of total production and remained largely intra-European until the late 19th century, government interference in agricultural trade has a long history. For example, in his Enquiry into the Nature and Causes of the Wealth of Nations, Adam Smith (1776) discusses at length the implications of tariffs in the agricultural sector of Britain, the European country *par excellence* where there has historically been a strong body of *laissez-faire* opinion. Smith's major concern was to conceive economic policies that would best promote the wealth of a nation, hence his advocacy of the *laissez-faire* dogma. Tariffs and protection were still commonplace in Europe at the time, partly a legacy of *mercantilist* thinking. Mercantilism, the dominant economic doctrine in Europe during the 16th and 17th centuries, mirrored the socio-political as well as intellectual developments of that time. Feudalism, characterized by a myriad of small and self-sufficient estates, was gradually giving way to strong nation states throughout Europe. The only way these nation states could grow was by means of international trade stimulation. However, according to mercantilist thinking, trade was subject to a striking dichotomy. The mercantilists advocated an export-led growth policy, coupled with restrained imports. A nation had to encourage strongly domestic production and exports, and was to discourage imports through tariffs, quotas, taxes, subsidies and other restrictions. Government interference was not confined to foreign trade regulation only; the stimulation of domestic production as well as of consumption of certain designated products was to be part of an overall government policy. For example, in 17th century Britain, the consumption of woolen goods was encouraged as a means to stimulate the domestic industry. From the late 17th century onwards, criticisms of mercantilism by prominent 'moral philosophers' such as the British Sir Dudley North and David Hume, the French Boisguillebert, and the Irish Richard Cantillon, paved the way to the triumph of the *laissez-faire, laissez-passer* doctrine, an intellectual construct of the Physiocrats in the 18th century, and embraced later in the teachings of Adam Smith. Combined with new ideas of individual freedom and universalism conveyed at the time through the Enlightenment, the new economic thinking inspired a generalized movement of regional economic integration in Europe. Between 1818 and 1924, 16 Customs Unions were formed in Europe.[3] Probably the most notable Customs Union created during that time was that of the German Zollverein, established in 1834 under Prussian leadership. This Customs Union implied the abolition of all duties at inner borders, and a duty levied on the outer frontier. Great Britain, motherland of *laissez-faire,* reduced gradually its customs duties in all sectors, from the early 1840s. The repeal of the Corn Laws in 1846 represented the most visible success of the new doctrine in the area of European agriculture. Consequently, Britain sought allies in Europe, who would agree to dismantle their protectionist policies. In particular, Britain signed the Treaty of Commerce with France in 1860, which led to a substantial reduction in French duties. Great Britain requested also increasing free trade with the 'new world' (today's North America). A relatively high level of protection in the USA at the time was explained by the high level of wages in the nascent American industries.[4] In Europe, Zollverein's duties on grain were abolished in 1853; duties on grain were abolished in The Netherlands in 1862, and foodstuffs could enter Belgium freely in 1871 (Tracy, 1989).

However, free trade in Europe was only short lived, for protectionism came back during the second half of the 19th century. This must be viewed as a combined response to large imports of cheap grain from the *new world,* as well as to the revival of nationalism in Europe, typified by the Franco–Prussian war of 1870–71. As agricultural prices started to embark upon a declining trend, from 1873 until 1896, protectionism gained gradually momentum in every economy of Europe with the exceptions of Britain, Denmark, The Netherlands, and to a lesser extent Belgium, where it remained quite moderate (Tracy, 1989). For example in France, the Méline Tariff of 1892 increased protection substantially in both the manufacturing and agricultural sectors, whereas in Germany a strong protectionist movement was backed up by Friedrich List's school of economic thinking (List, 1841). However, a small respite in German protectionism took place during the last decade of the 19th century. After Bismarck's dismissal in 1890, Germany, which was keen to expand its export markets, concluded a series of new commercial treaties with Austria, Hungary, Italy, Switzerland and Belgium, involving a decrease in import duties on a number of food commodities.[5] A Free Trade school started to appear during the first decade of the 20th century in Germany, but it had only a marginal representation in the *Reichstag*.

3.3.2 The inter-war period (1918–39)

Protectionism relaxed somewhat at the turn of the century, until tariffs were generally phased out during the mid-1910s. However, a new wave of protectionism started during the 1920s. Serious price fluctuations, indebtedness, and the advent of the Great Depression led many governments to use trade protection again as the main cornerstone of their agricultural policy. Government intervention in European agricultural markets grew enormously during the 1930s.

The Depression, characterized by declining prices and food shortages, was particularly acute in Germany where the Treaty of Versailles had suspended many of the country's economic rights. However, in the mid-1920s, Germany was able to recover the right of formulating an independent commercial policy. Exports of agricultural commodities were permitted again. Tariffs on agricultural produce were reintroduced with the aim of halting the fall in prices in 1925. Although the policy succeeded in restraining imports of foodstuffs, it did not unfortunately prevent the large fall in agricultural prices after 1929. The many indebted farmers were ruined and this facilitated the rise of Nazism in the 1930s. The Third Reich defined a clear policy for agriculture. Darré, who was in charge of a new agrarian policy for Germany at the time, did not see the farmer as an ordinary entrepreneur, and he could not conceive that the production of food be subjected to the free play of market forces (Tracy, 1989). Consequently, the essential aims of the National Socialists were fair prices to the farmers and food self-sufficiency. In 1933, the Nazi subjected almost all agricultural imports to the control of state import boards,[6] and they were among the first in Europe to devise elements of a structural policy in the sphere of agriculture, in allocating financial aid primarily to the viable farms. All these elements were combined in a coherent policy that was systematically put into practice and that resulted in heavy centralization.

Heavy centralization taking the form of a strict regulation of agricultural markets and in line with an overall economic plan, was also the dominant feature in another totalitarian economy at the time: Italy in the 1920s. Shortly after Mussolini seized power in 1922, agriculture became an essential chapter of the fascist government's economic policy, given the important rural component of the fascist electorate. An agricultural policy centered on the objective of self-sufficiency was initiated. The new laws of 1923 and 1928 introduced the policy of *bonifica integrale* (integral land improvement) which turned any unused land, as well as some areas contaminated with malaria, into viable and cultivated agricultural areas. The social objective of the new policy was to favor small farmers (Balcet, 1997). Import substitution helped reduce the magnitude of trade deficits. In order to achieve the objective of self-sufficiency, Mussolini launched the famous *Bataglia del Grano* (the Battle for Grain) spanning over the years 1922–29. Great emphasis was placed in particular on wheat production. Wheat yields were to be raised with the help of subsidies, of import duties as well as of appropriate technical advice.[7] In the mid-1930s, a monopolistic structure was created in the wheat sector, enabling wheat growers to deliver their production to a central organization at fixed prices. The type of production favored by the government at the time corresponded to these agricultural products in excess supply on world markets. The results of this interventionist and protectionist policy fell short of the expected objectives. In spite of an increase in the domestic production of cereals since the early 1920s, a third of cereals consumption was still met by imports in 1937, mostly from Argentina (Balcet, 1997).

Although a non-totalitarian economy, France also resorted to trade protection after the Great Depression. Import quotas were introduced in the 1930s on almost all commodities with the exception of wheat, which was subject to import duties, buffer stocks, as well as to other measures aimed at stimulating domestic consumption.[8] Wheat was indeed given a particular consideration in the 1936 Léon Blum's Government of the *Front Populaire*. The Popular Front created the *Office National Interprofessionnel du Blé* (ONIB) which was to regulate all foreign trade and to fix wheat prices. The number of farm organizations proliferated in France during that time, as specialist producers' organizations appeared for example in the wine and sugar beet sectors. The *Chambres d'Agriculture*, created at the level of each *département*, were operating from 1927.[9] The French farming community was becoming increasingly politically organized. French farmers, and indeed most European farmers, started taking a very active role in the class war that was appearing in late 19th century Europe, and that asserted itself after the First World War (Duby and Wallon, 1976). Consequently, farmers' associations started exerting an increasing pressure on the governments all over Europe, and they became an essential ingredient in the formulation of price support policies in Europe.

Spain is a country with a long-established protectionist tradition. Between 1760 and 1900, the doubling of the Spanish population to 18.6 million was made possible without an increase in the dependence *vis-à-vis* food imports, in spite of extremely low productivity gains in the agricultural sector over the period. Imports were limited by means of extremely high import duties. For example after the 1880s, import duties on wheat were among the highest in Europe (Simpson, 1995). From the end of the 19th century onwards, Spain started to establish its comparative advantage in the wine, olive oil as well as citrus fruit sectors. The First World War was accompanied by a generalized movement of protection in the Spanish agricultural sector, and price support policies became the major ingredient of the new policy. Agricultural prices were being fixed, and maximum target prices, as well as minimum guaranteed prices were becoming the norm. Moreover, import quotas were introduced in 1921, in addition to the already high existing tariffs. As a result of this protectionist policy, Spain's self-sufficiency rate in wheat was close to 100 per cent at the end of the First World War (Simpson, 1995). In 1936, the market for wheat was further organized and centralized through the creation of a state controlled trade board. The beginning of the 20th century was witness to substantial improvements in Spanish agricultural productivity. These favorable developments were nevertheless halted by the Spanish civil war (1936–39), to such an extent that, at the end of the 1940s, the pre-war level of agricultural output had not yet been attained. According to Simpson (1995), the Spanish economy had to wait until 1954 before the 1929 level of agricultural output was to be reached again. The results of many decades of agricultural protection in the case of Spain have been noticeable at the structural level. Price support policies have retarded and prevented structural change by slowing down the exit of the labor resource from the agricultural sector. In that regard, Spain is only one of the many examples of a country where structural change has been hampered by protectionism. Yet, the case of Spain is unique in that the level of protection before the Second World War was the highest in Europe. Spain was also the country in Europe that displayed the lowest levels of productivity in the agricultural sector at the time. It appears that the high level of protection was in part a direct consequence of the political pressure exercised by the then very large farmers (Simpson, 1995). Not surprisingly, Spain's protectionist policy favored mostly the rich farmers, producers of cereals at the expense of the poor majority rural population. The policy also had devastating effects on the level of prices for Spanish consumers.

In the newly created Irish state, the *Fianna Fáil* party took office in 1932 with a definite protectionist policy aimed at reducing economic dependence on Britain, a nationalistic objective, and at providing employment, thereby reducing emigration, a socio-economic objective (Kennedy *et al.*, 1988). The agricultural sector was at the heart of de Valera's new government's economic policy. The year 1932 also marked the beginning of a 'tariff war' between Ireland and the UK, which was to last for six years. This was initiated by the inability of both Irish and British governments to find a compromise in relation to the payment of annuities by Irish tenants acquiring land and farms from British landlords since the late 19th century. The transfer of land and farms to Irish tenants had been made possible by a number of schemes funded by the British government. One such successful land transfer scheme was the 1885 Ashbourne Act, and it permitted Irish tenants to pay an annuity over a period of time to the British government that had once advanced the purchase money to the British landowners (O'Malley, 1982). In de Valera's view, the annuities still payable after he came to power were to be payable to the Irish government, as opposed to the British government in the past. De Valera's strong refusal to transfer the receipts of land annuities to the British government, led Britain to initiate a number of economic sanctions affecting Ireland's agricultural sector. *Ad valorem* custom duties (of 20 per cent) were imposed on Irish agricultural products imported in Britain, and import quotas were introduced in 1934. These quotas effectively meant a total prohibition of Irish veal and beef on the British market at the time. Ireland's immediate reaction was to impose heavy custom duties on manufactured imports from the UK, in particular, in the cement, coal and steel sectors. Subsequently, *ad valorem* tariffs were introduced in both agricultural and manufacturing markets, ranging from 15 to 75 per cent (O'Malley, 1989). Protectionism in Ireland should be seen partly as a response to the British sanctions following the annuities dispute mentioned above, and partly as a deliberate strategy enshrined in a new economic logic of greater economic independence *vis-à-vis* the British market. Self-sufficiency in the agricultural sector was a prime objective at the time, and subsidies were granted to producers of various products such as wheat and linen, so as to increase self-sufficiency rates in these sectors.

Protectionism appeared even in the traditionally more market-oriented economies of Europe such as Denmark and Britain. When protectionism came back in Europe, Denmark was able to adapt to the new world-trade conditions by specializing in livestock production at the expense of cereals production, an increasingly protected sector. However, in the early 1930s, Danish exports of livestock products became severely affected by the fall in purchasing power abroad, and also by the increased import restrictions applied in other European countries. Since Danish agriculture was depending to a great extent on trade, immediate efforts to sustain positive trade relations with Denmark's main European partners were actively pursued by the Danish government. A short-lived Anglo-Danish commercial Treaty was signed in 1933, under which Britain undertook not to impose duties on Danish bacon and ham, and a bilateral agreement was signed with Germany in 1934 (Tracy, 1989). However, in signing these two commercial treaties, Denmark found itself in a weak position. For example, the increase in Danish exports in Germany was subject to Denmark raising its imports of German foodstuff by an equivalent amount. This proved increasingly difficult for Denmark as it started to face a shortage of foreign exchange. The condition imposed by Germany became *de facto* a subtle import restriction, and overproduction combined with declining prices called for government intervention in this sector. Substantial elements of protection were also introduced in the sugar beet sector; variable levies were imposed on butter in order to maintain a minimum price; a very advanced and successful pig scheme aimed at cutting drastically the production of pigs in the country, and at raising its per unit price, was devised in 1933.[10]

In the context of falling world agricultural prices in 1929 and afterwards, and because of the resurgence of protectionism at the world level, the British market was temporarily used as an 'absorber' of agricultural surpluses. Departure from free trade principles became inevitable, and Britain started to discriminate against foreign (non-Dominion) foodstuffs. Its policy was in essence defensive and reactive; for example, Britain reacted to the Irish Prime Minister's nationalistic policy by imposing duties of 20 per cent on all main agricultural imports from the Irish Republic. Government intervention in British agriculture must be seen as a solution of last resort. Protectionism and intervention were ultimately devised as the last possible solutions to the problem of falling agricultural prices and incomes during the 1920s, after many attempts at solving the problem through free trade policies. The governments at the time (in particular the labor governments) became also more and more accepting of the plight of the farming community. The principle of agricultural support in Britain was finally accepted by the 1939 government (Smith, 1990). Free trade principles gave gradually way to intervention and support mechanisms, such as customs duties, quotas, and so on. The 1947 Agriculture Act laid down the foundations of post-war agricultural policy in the UK. Agricultural production was stimulated through agricultural expansion programs, based on deficiency payments systems.

It is clear that during the inter-war period, protectionism made huge advances in all countries of Europe albeit with varying degrees, and in the rest of the world. However, nowhere else as in Germany was the agricultural policy so clearly defined and so comprehensive. A common core to all these agricultural policies was their emphasis on trade limitations with import restrictions and export subsidies as the major forms of intervention. Moreover, there was a growing awareness during that time that the agricultural sector was different from any other sector of the economy, and that it was much more subject to large price fluctuations and to long-term market dis-equilibria than any manufacturing sector in any economy. This awareness called for additional forms of market intervention; this is how the first marketing schemes were created in both Denmark and Britain.

The long-term implications of protectionism have often been decried. Farmers gain temporarily as they become shielded from the effects of depression, but in the long term protectionism tends to halt positive structural adjustments in the agricultural sector. The need to devise immediate solutions in a context of generalized crisis led pre-Second World War European governments to overemphasize price support policies, and to divert their attention away from any structural measures. In general, European governments failed to encourage technological progress, to consolidate fragmented agricultural units and to develop infrastructure in the countryside. However, we have shown that protectionism has been introduced with varying degrees of intensity and at different times in the various European countries before the Second World War. For example, as they held to the *laissez-faire* doctrine before the First World War, Britain, Denmark and The Netherlands were able to promote substantial structural adjustments in their agricultural sectors.

In Britain for example, the arable surface declined, non-viable farms disappeared, and the large units were able to achieve high productivity rates. Although a *laissez-faire* approach is undoubtedly more conducive to enabling structural change than interventionism, the relationship between positive structural adjustments and lack of government intervention should however not be exaggerated. As we will see in Chapter 5, Dutch farmers were able to establish themselves as the most competitive farmers in Europe, thanks mostly to a radical change in attitude with regard to innovation; this was very much supported by government-led research and educational schemes. What seems therefore to matter is more the type of government interference than intervention *per se*.

By contrast, Mediterranean countries – and in particular France, Italy and Spain – tend to lie at the other end of the spectrum. Very little was done until the 1950s to promote technical change, and these countries had, with the exception of Germany, the most protected agricultural sectors in Europe. As a result, the structure of their protectionist agricultural sectors remained virtually unchanged. For example, and in spite of productivity gains achieved during the 19th century, French agriculture was still lagging well behind that of other European countries such as Belgium, Sweden, Switzerland and Germany. In the beginning of the 20th century, the production of calories per unit of labor in France was well below that of north-European countries, and in spite of its many trade barriers, French agriculture was still unable to produce sufficient output to meet domestic demand (Duby and Wallon, 1976).

3.3.3 Post-war recovery

After the Second World War, maximization of food production became the predominant preoccupation throughout Europe. There was also a general need to save foreign exchange by keeping imports as low as possible. In order to raise production as fast as possible, state involvement increased rapidly in all the countries of Europe. In particular:

- price support policies were multiplied or reinforced,
- income guarantees were given to farmers, and
- farm investment and improved farming methods were encouraged by credits and subsidies.

Economic recovery was greatly stimulated by American aid allocated through the Marshall Plan in 1948 and afterwards. In the late 1940s, agricultural production was back to its pre-war level in Europe, whereas by the end of the 1950s, it was 50 per cent above it, for a population that had increased by roughly only 20 per cent. The impressive growth in production went in parallel with a large fall in the numbers employed in the farming sector. Between 1950 and 1960, the agricultural labor force in the OEEC area[11] decreased by 15 per cent (Tracy, 1989). Economic recovery, connected with the necessity to rebuild the devastated European economies, offered many alternative opportunities to the rural population. Urban centers grew enormously, essentially in the countries that had little *laissez-faire* tradition, with the manufacturing sector representing the major and the most attractive growth area. As agricultural supply caught up with demand in the early 1950s, the objective was no longer to increase production at all costs, but to raise agricultural efficiency. The widespread introduction of tractors, the increased use of fertilizers and the diffusion of technical knowledge, all helped increase yields of crops and livestock. Labor productivity increased faster in agriculture than in other sectors, in every European country except Ireland. This was a country still relatively sheltered from economic forces infused from mainland Europe, and structural adjustment in the Irish agricultural sector was thus minimum during that decade. A change in emphasis finally occurred in 1958 when the 'export-led growth' philosophy, based on industrialization-by-invitation became the new strategy of the various Irish governments. Shortly after the Second World War, the agricultural sectors of Greece, Spain and Portugal also remained isolated from international market forces, due to their isolationist policies. In these three countries, the change in economic and political philosophy preceded EC membership by a relatively short period.

At the same time, concern with the relatively low level of farm incomes became more acute. Incomes and living standards on farms did increase in absolute terms, partly because of the large outflow of labor from the sector during the 1950s. But, because income elasticity of demand increasingly fell, agriculture's share in national income was actually falling, and the transfer of manpower from agriculture to other sectors was not sufficient to avoid a further widening of the income gap between agriculture and the rest of the economy. Therefore, agricultural policies tended to become increasingly costly and complicated, as the farm organizations increased greatly in strength. These farm organizations accentuated their demands on price support, although movement out of agriculture would have been the only way to raise income per head in the long run. From the standpoint of pure economic theory, the governments of Europe should have encouraged the trend of falling agricultural employment, but most governments were reluctant to offend the farm organizations and their policies continued to emphasize price support. This is in this broad context that the Treaty of Rome was signed and that the CAP was born.

Summary

Agricultural support policies and structural policies are the two broad types of policies affecting the agricultural sector in any country. Whereas agricultural support policies are aimed at stabilizing prices and farmers' incomes, structural policies stimulate the change in the economic structure of a given country, that is they affect the relative importance of the various industries therein. The rationale for government intervention in agriculture stems from the specific economic nature of the agricultural sector. Supply is highly unpredictable, because of natural phenomena, and demand is not very responsive to price and income changes. In the case of the EU and of many other economies of the world, government intervention in the agricultural sector is also explained by both socio-political and strategic considerations. In Europe generally, agricultural support policies have tended to be more common and prominent than structural policies. Various instruments applied at the farm, domestic and frontier level have had different impacts on price, income, demand and supply levels. The various impacts are studied with reference to an ideal situation where competitive pressures prevail in all other national and international markets. It was shown that all policy instruments tend to stimulate domestic production by penalizing imports from the rest of the world, as well as either domestic consumers (as in the case of import quotas and variable levies) or domestic taxpayers (as in that of deficiency payments). Also, some of these policies tend to depress world prices.

The CAP did not create government intervention in Europe, for there was a long tradition of government intervention in European agricultural sectors prior to the formation of the Common Market in the 1950s. For example, in spite of its long-standing *laissez-faire* attitude, and faced with increasing protectionism worldwide, the UK resorted to accepting the principle of agricultural support shortly before the outbreak of the Second World War. Pre-war government intervention and protectionism in agriculture were most intense in the then totalitarian economies of Germany, Italy and Spain. However, in other countries such as France, the imports of cheap grain from the new world in the late 19th century had also precipitated an arsenal of protective measures in favor of domestic farmers before the Second World War. Finally, another specific feature of pre-war agricultural markets has been the increasing influence of politically organized farming communities across Europe. Farmers' associations started asserting themselves as an increasing influential actor in the determination of price support policies in Europe. These associations tended to emphasize price and income support policies, although it was obvious at the time that a re-allocation of production factors, and in particular of labor, out of agriculture was the only economically viable way to raise income per head in the long run.

Key terms and concepts

Price-support policies
Structural policies
Price stabilization
Deficiency payments
Tariff protection
Variable levies
Import quotas
Government intervention
Protectionism
Farmers' associations

Notes

1 Thanks to Joseph's accurate interpretation of the Egyptian Pharaoh's dreams, a severe famine could be avoided in the land of Egypt by piling up and setting aside stocks of grain and food during seven years of abundant food supplies.

2 Note that there are some unusual circumstances in which the imposition of a tariff can reduce the domestic price of the imported good; this is known as the Metzler Paradox (Metzler, 1949).

3 Jacob Viner (1950) defines a Customs Union as a grouping of member nations that meets the following conditions: (i) complete elimination of tariffs as between the member territories, (ii) establishment of a uniform tariff on imports from outside the Union, and (iii) apportionment of customs revenue in accordance with an agreed formula.

4 The decrease in protectionism in the 1840s in the USA was only short lived.

5 For more on this, see Tracy (1989: 89).

6 The so-called *Reichsstellen*. In addition, other boards were set up such as the *Reichsnährstand*, with the aim of organizing and regulating all aspects of food production and distribution at the national level.

7 Note that the Battle for Grain policy stimulated the production of wheat at the expense of other food commodities. In particular, the Italian economy suffered from acute shortages in the livestock products sector.
8 Such as the imposition of a milling ratio which required millers to use a large proportion of domestically produced wheat (up to 97 per cent) in their flour.
9 Note that in the early 1930s, the interests of French farmers were represented in a specific branch of the Socialist political party of the time: this was called the *Confédération Nationale Paysanne*. It dissolved after the fall of the Popular Front in 1938.
10 According to Tracy (1989), this scheme was probably the first marketing quota system ever introduced.
11 Organization for European Economic Co-operation, which was established in 1948 to administer American Aid under the Marshall Plan. It later expanded to include non-West-European countries, and it became the OECD.

Bibliography

Balcet, G. (1997) *L'Economia Italiana – Evoluzione, problemi e paradossi* (Feltrinelli: Milano).

Colman, D. and Young, T. (1993) *Principles of Agricultural Economics – Markets and Prices in Less Developed Countries*. (Cambridge: Cambridge University Press).

Duby, G. and et Wallon, A. (1976) *Histoire de la France Rurale* (Tome 3, de 1787 à 1914), Editions du Seuil: Paris.

Kennedy, A.K., Giblin, T. and McHugh, D. (1988) *The Economic Development of Ireland in the Twentieth Century* (London: Routledge).

Krueger, A.O., Schiff, M. and Valdés, A. (1988) 'Agricultural incentives in developing countries: measuring the effect of sectoral and economy-wide policies', *World Bank Economic Review*, 2, No. 3, 255–71.

List, F. (1841) *National System of Political Economy* (London: Dent).

Metzler, L.A. (1949) 'Tariffs, the terms of trade, and the distribution of national income', *Journal of Political Economy*, 57, 1–29.

OECD (1995) *Agricultural Policies, Markets and Trade in OECD Countries. Monitoring and Outlook*. Paris.

O'Malley, L. (1982) *Business Law* (London: Sweet and Maxwell).

O'Malley, E. (1989) *Industry and Economic Development: The Challenge for the Latecomer* (Dublin: Gill and MacMillan).

Simpson, J. (1995) *Spanish Agriculture – The long siesta 1765–65*. (Cambridge: Cambridge University Press).

Smith, A. (1776) *An Enquiry into the Nature and Causes of the Wealth of Nations* (Reprinted 1976, Chicago: Chicago University Press).

Smith, M.J. (1990) *The Politics of Agricultural Support in Britain – The development of the Agricultural Policy Community* (Dartmouth: Aldershot).

Tracy, M. (1989) *Government and Agriculture in Western Europe 1880–1988* (3rd edition, New York, London: Harvester Wheatsheaf).

Viner, J. (1950) *The Customs Union Issue* (London: Stevens and Sons).

4
The Common Agricultural Policy

Objectives of this chapter

This chapter is intended at:

- Describing the birth, principles, policies and development of the Common Agricultural Policy (CAP).
- Analyzing the relative success/failure of the policy.
- Studying the progression towards the 1992 CAP reform, as well as further reform processes.

Introduction

After the Second World War, the *rapprochement* between France and Germany became a priority for many West European statesmen. The only way a new conflict could be avoided between the two belligerents was through the construction of an economically and politically integrated area in Europe. This was espousing a view that had been expressed by Vilfredo Pareto at a Peace Congress in Rome in 1889 (Pareto, 1889). Pareto argued that customs unions and other international economic arrangements were a secure route to better political relations and to pacification. Schuman's declaration of May 1950 was aimed at bringing a 'durable solution to the German problem', and it underlined the decisive role of the binomial France–Germany. Subsequently, the concept of a 'European Community' appeared with the signing in 1951 of the Treaty of Paris, which created the European Coal and Steel Community (ECSC). This treaty was in fact a clear plan to make another war materially impossible, as the French and German war industries were removed from national control. Because of harsh difficulties encountered by both the German Chancellor Adenauer and the French President Coty in the definition of a political regrouping, the following treaty, ratified in Rome in March 1957 and establishing the European Economic Community (EEC), was subsequently limited to economic aspects. Nowhere else than in the agricultural sector have the principles of economic integration been applied so fully in the nascent European Common Market of the 1950s.

4.1 Birth and mechanisms of the CAP

The rationale for creating a sophisticated organization of agricultural markets at the European level can be summarized by three reasons:

(i) *Political reasons*: The food shortages endured during wartime did show that agricultural products could be used as a form of weapon. They demonstrated clearly the strategic role of agriculture in the various economies.

(ii) *Social reasons*: The importance of the farming population, representing still a quarter of total civilian employment in France, a third in Italy, and 17 per cent in Luxembourg in 1960, could not be neglected. This socio-economic group was well known for representing the backbone of conservative (anti-leftwing) parties in Europe at the time of the cold war. In particular, the importance of *le vote paysan* in France (the peasants' vote) has been acknowledged as a balancing force in French politics until the election of President Valery Giscard d'Estaing in 1974.

(iii) *Economic reasons*: The unpredictability of agricultural supply, and the consumer's shift away from food products, all aspects that have been discussed in Chapter 3.

In addition, other problems had to be solved such as the disparities in food prices between member countries, and their distortion effects upon industrial costs and wages. The six founding members unanimously accepted that agriculture be organized in common. Indeed, some countries had similar objectives: the French, Dutch and Italians,[1] the major producers, were keen to open their markets to West German manufactured goods in return for free access to the large and rapidly expanding German market. Although there was consensus on the principle of organizing agriculture into a common and unified market, many conflicts arose as to the type of instruments to be employed, as well as to the degree of support to be introduced. On the one hand, the efficient

Dutch producers, processors and traders, wanted as free a market as possible, and on the other, the French, who possessed at the time 50 per cent of the agricultural area in the Community, insisted on a market organization which would ensure adequate returns and preference over imports from non-EEC countries (Tracy, 1989). Both countries were major producers in the Community. The Dutch were particularly strong in the dairy sector, and the French in the cereals and sugar beet sectors. Consequently Dutch and French made their voice heard strongly, and as a result, the form of the CAP was largely due to them. Indeed, dairy products, cereals and sugar beet profited the most from the CAP. The other countries, Belgium, Germany, Luxembourg and Italy, as net importers were somewhat reluctant to open their import markets to their partners; however, as high price countries (Italy excepted), they could not allow the real income of their farmers to be undercut. They had little choice but to accept the architecture of the CAP.

In spite of all these conflicts, a first report encompassing some provisions relating to agriculture had to be written, which reflected all these ambiguities. This report, known as the Spaak Report, was approved by the foreign ministers of the six founding EEC countries in Venice in May 1956. The Spaak Report formed the basis of the Treaties establishing the EEC (Treaty of Rome) and the EURATOM (European Atomic Energy Community). Signed in Rome on 25 March 1957, the two treaties entered into force on 1 January 1958, after ratification by the six national parliaments. The Treaty of Rome provided for the establishment of a Common Market, in which custom duties and quantitative restrictions on trade between member states would be gradually removed and a common external tariff established. Whereas article 2 (Part 1) of the Rome Treaty sets up the objectives of the EEC, article 3 clarifies the type of activities that should lead to the attainment of such objectives. It states in particular that:

> the activities of the Community shall include, on the conditions and in accordance with the time-table provided in this Treaty [...] the establishment of a common policy in the sphere of agriculture.

Although transport was another area that was accorded priority by the Community, in no other sector of the various countries was economic integration planned to be as advanced as in the agricultural sector. The free trade arrangements envisaged by the Treaty of Rome were to be applied gradually in every goods market, but in addition the agricultural markets of the six founding members were to achieve a high level of economic integration. Not only had tariffs on trade between member states to be removed over the transitional period (of twelve to fifteen years), and a common external tariff to be imposed, but in addition, a common organization of markets was to be established in the agricultural sector alone, by the end of the transitional period.

A Common market organization for European agriculture means by definition that:

(i) all obstacles to trade of agricultural products as between member countries must be removed,
(ii) for any given commodity produced in the Community, a unique price needs to be established and maintained,
(iii) the policy must be funded by a Common budget, and must be administered in a concerted manner.

An important principle underlying criteria (i) is that of *community preference,* which entails that the economically integrated area must protect itself from external influences and threats. The progress towards a Common Market in agricultural products was to be gradual.

Ten articles in the Treaty of Rome deal with the CAP (articles 38 to 47, Part II, Title II). In particular, article 39 (renumbered as article 33), which is probably the most well known, sets the policy objectives of the CAP, for example:

(i) to increase agricultural productivity through encouraging technical progress and the optimal utilization of factors of production, especially labor,
(ii) to achieve a fair standard of living for the agricultural population, with increases occurring in the individual earnings of persons engaged in agriculture,
(iii) to stabilize markets,
(iv) to guarantee regular supplies,
(v) to ensure reasonable prices to consumers.

Article 39 is only a statement of objectives, opened to many different and contradictory interpretations. Listing the increase in agricultural productivity through technical change and through the optimal utilization of production factors as the number one objective would suggest that the policy would favor structural measures rather than overall price support. In order to elaborate further on these objectives, as laid down in the Treaty of Rome, and on the means of their implementation, a conference of agricultural ministers was held in Stresa in July 1958. At the conference, a number of resolutions were passed, dealing with:

- the expansion of intra/extra Community agricultural trade,
- elements of structural policy,
- the elimination of quantitative restrictions.

However, it failed to give clear indications on the future instruments of the gestating CAP. After many negotiations, proposals, draft regulations, consultations and Council sessions, a package establishing the common methods of support for the most quantitatively important agricultural products of the six founding members was finally adopted on 14 January 1962 (Tracy, 1989). Although a possible interpretation of article 39 could have suggested that preference be given to the increase in agricultural productivity and therefore, to structural measures, in subsequent practice not much attention was paid to structural measures. The preference went to agricultural price support. In 1960, the Commission had already proposed a system of agricultural price support based on variable import levies and on market intervention. An agreement on these principles was reached at the end of 1961 for cereals, poultry, eggs, pig meat, fruits and vegetables. Two years later, other commodities were included in the scheme: beef, veal, dairy products and vegetable oil.

4.1.1 The mechanisms of the CAP

Prices in the EU have been maintained for many decades at a relatively high level through various protective measures. Based on overall market price support, the policy presented at its inception four major components:

(i) a target price,
(ii) an intervention price,
(iii) export refunds or restitution payments,
(iv) the 'green money', and the monetary compensatory amounts (MCAs).

(i) Target price

Target prices are fixed on a yearly basis by the EU Council of Ministers. This price, shown as P_{ta} in Figure 4.1 is the highest possible desirable price for a given product quality, or the optimum price that is expected to be achieved given the CAP objectives. It is in fact a reference price for the wholesale market. Although being a common tool, target prices are varied regionally in order to take account of transport costs between

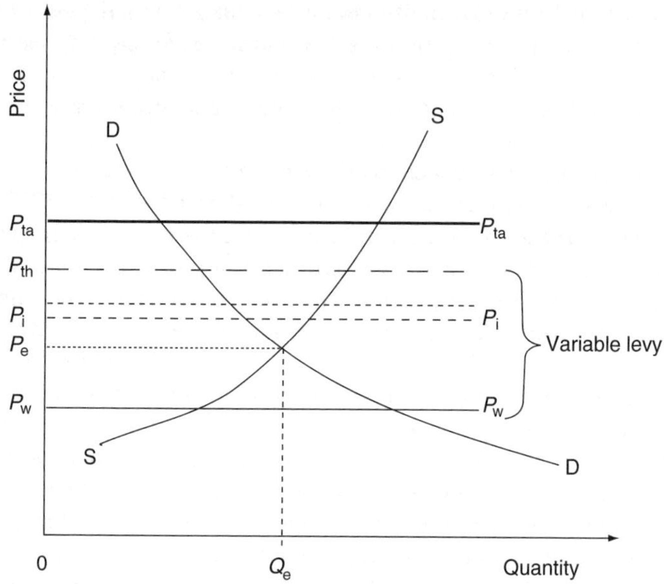

Figure 4.1 CAP price mechanisms

producing and consuming areas, as well as storage costs. Hence the target price is a bit more than a producer price.

When transport and storage costs are subtracted from P_{ta} the resulting price is the threshold price P_{th}. Given the structure and level of efficiency of the agricultural holdings in the EU, the target price (and hence the threshold price) for a given commodity is normally well above the world price (P_w in Figure 4.1). Target and threshold prices are maintained through import controls. In order to keep the EU farming sector in isolation from the rest of the world, variable levies are charged on all imports entering the EU, essentially through the port of Rotterdam. The levies are variable because they are calculated on a daily basis by the EU Commission as the difference between the threshold price P_{th} and the world price P_w, corresponding to the *cif* (Cost Insurance Freight) price at Rotterdam.

The implications of this policy for the domestic (that is EU) consumers, producers and for the rest of the world are easy to understand. In Figure 4.1, it can be seen that if imports were free to enter the EU market, the equilibrium market price would stabilize at P_e and the quantity sold at Q_e. Assume that a target and a threshold price are set at a level above equilibrium, as is normally the case. In particular, the diagram shows the threshold price P_{th}. The EU producers see this price as an incentive to increase their level of output. On the other hand, a higher price would be a disincentive for the EU consumer. Consequently domestic production raises whereas domestic consumption declines. In order to restore the balance, a variable levy ($P_{th} - P_w$) must be imposed. The European producers enjoy a bigger market and a higher price, at the expense of foreign producers who are penalized by a contracted EU market. Excess of supply at the world level would translate through lower international prices. Because the EU is a large agricultural producer and trader in the world, the policy would tend to depress world prices (see Chapter 3).

(ii) The intervention price

Whereas the target price is the upper limit or optimum price that should be attained on every market on a daily basis, the CAP allows the actual market price to oscillate within a permissible band of fluctuations. The lower limit of this band is known as the intervention price (P_i in Figure 4.1). Target prices are used as a calculation basis for determining at which price level the intervention takes place. The intervention price is usually 5 to 10 per cent below the target price. If the market price falls to the intervention level, produce is purchased for storage by an intervention agency. Thus, the intervention price is virtually a minimum guaranteed price. The quantity bought by the European intervention agency is used as buffer stock in order to regulate the market and to maintain the market price within the permissible band of fluctuations.

(iii) Export refunds or 'restitution payments'

If the policy had been limited to the provisions described above, virtually all EU exports of agricultural commodities would have ceased. In order to encourage exports of excess supply to the rest of the world, provision has been made for the payment of export subsidies, euphemistically termed 'restitution payments'. At any given world price P_w, the export refund is at least equal to the minimum ($P_i - P_w$) and to the maximum ($P_t - P_w$).

As a result, European producers can export agricultural commodities at profit even though intervention prices (that is prices held within the Community) are above world market prices.

(iv) The 'green money' and the MCAs

From the onset, the agricultural prices fixed by the Council were denominated in a European Unit of Account (EUA). All transactions of the then EEC, as recorded in the EEC budget, started to be recorded in EUA terms.

At the very beginning, the EEC used the Belgian Franc as the common denominator; it is only in 1961 that it adopted a unit of account based on a gold parity.[2] Shortly after the collapse of the Bretton Woods system, the unit of account became linked to the 'joint float' in the 'monetary snake' system. The negotiation of the Lomé Convention in 1975 led to an important change in the European unit of account system. Following a proposal put forward by the French, the Commission adopted the basket unit of account (Strasser, 1992). The basket, often referred to as a 'cocktail of currencies', is a weighted sum of the various Community currencies, that is:

$$EUA = \sum_{i=1}^{9} (m_i \cdot w_i)$$

where:
m_i is a given member state's currency (that is Deutsch Mark, French Franc, Italian Lira, and so on), and
w_i, a weighting coefficient calculated as follows:

$$w_i = r \cdot s_i$$

where:
r is the *central rate*, that is the exchange rate between any nine currency and the EUA, and

$$S_i = f(GNP, X + M, ms)$$

The share s_i is thus calculated on the basis of a five-year average of a member country's GNP,[3] the level of this country's intra-EU trade, as well as its participation in short-term monetary support (*ms*). The share s_i reveals the economic importance of a given member state in the community. For example, the shares of the German and French economies were initially 27.3 and 19.5 per cent respectively. The share of Luxembourg, a wealthy but minuscule economy, was only 0.3 per cent.

Initially the basket unit of account was made up of the following amounts of member states' currencies:

$$\begin{aligned} \text{EUA} = {} & 0.828 \text{ DM} + 1.15 \text{ FF} + 0.0885 \text{ UK£} + 109 \text{ ITl} \\ & + 0.286 \text{ NLgl} + 3.66 \text{ BF} + 0.217 \text{ DKr} + 0.00759 \text{ IR£} + 0.14 \text{ LF} \end{aligned}$$

where:

DM Deutsch Mark,
FF French Franc,
ITl Italian Lira,

NLgl Dutch Guilder,
BF Belgian Franc,
DKr Danish Krone,
IRE Irish Punt, and
LF Luxembourg Franc.

The EUA was designed as a fixed unit, which remain unchanged until the Community was enlarged to 12 members. When the European Monetary System started to be in operation in March 1979, the EUA was given a value equal to that of the well-known European Currency Unit (ECU). However, the specificity of the ECU was that its composition could be altered at least once every five years, or at any moment, if required.[4]

As long as the exchange rates remained stable in Europe, agricultural prices, denominated either in EUA, or later in ECUs, were unaltered and were compatible with the principle of price unity across member states. However, monetary disturbances started to be felt in the late 1960s in Europe. More particularly, political and social unrest in France, combined with looming economic difficulties, compelled the French authorities to devalue the French Franc by 11.11 per cent in August 1969. These events took place only two years after the common guaranteed prices were in full operation. Moreover, the loss of confidence in both sterling and the dollar nourished positive speculative pressures in favor of the DM. During the 1960s, a huge amount of dollars had flooded the European exchange markets. As European and Japanese firms were gaining in competitive ability relative to that of US firms, the US payments position shifted into large overall deficits. Soon after the devaluation of the French Franc, the Deutsche Mark was revalued by 9 per cent.

The principle of price unity within the Community was irremediably undermined. Two options were on offer to the then EC's Agri-Monetary Management Committee: readjust the parity of the Unit of Account, or neutralize the effects of both the devaluation and the revaluation through the introduction of a new policy. The first option would have led to a deadlock: since the FF and the DM moved in opposite directions, it was impossible to decide whether a readjustment of the EUA should have taken the form of a devaluation or of a revaluation. Therefore a new policy was introduced which offset the fear of inflation in France and of loss of revenue among German farmers: the intervention price was devalued in France, by the full amount of the devaluation so as to eliminate unfair gains to French farmers exporting elsewhere in the Community; conversely, the intervention price was revalued in

Germany so that the German farmers could keep the same standard of living as their other European counterparts. A system of border taxes and subsidies was born, known as the MCAs. This system is explained in Box 1.

Box 1 How did the MCAs work

1 Time period t_1 (pre-currency fluctuations)

Assume that the intervention price for a ton of wheat is EUA 100, and that the official or central exchange rates are:

$$\text{EUA } 1 = \text{FF } 6 = \text{DM } 2.5 \qquad \text{(eq. 1)}$$

This implies that a typical French farmer receives FF600 for each ton produced in t_1.

2 Post currency fluctuations

In t_2, after the devaluation of the FF and the revaluation of the DM, the new *central rates* are:

$$\text{EUA } 1 = \text{FF } 6.5 = \text{DM } 2 \qquad \text{(eq. 2)}$$

According to equation 2, the French farmer would receive FF 650 for each ton produced (and the German farmer DM 200 instead of DM250 as in t_1). Hence the French farmer would enjoy an unfair advantage compared with his other European counterparts, who still receive EUA 100 for each ton of wheat produced. Likewise, the German farmer would lose DM 50 for each ton produced.

Solution: introduce a 'green exchange rate', in order to keep 'green prices' unaltered. This 'green rate' is such that:

$$\text{EUA1} = \text{FF } 6 = \text{DM } 2.5 \text{ DM} \qquad \text{(eq. 3)} \quad \text{(note: same as eq. 1)}$$

Equation 3 implies that the intervention price in national currency has been devalued in France. The French farmer involved in cross-border trade receives FF 600 for each ton of wheat.

It follows that the MCAs are simply intra-Community border taxes and subsidies. In the case of France, they are calculated as follows:

$$\text{green rate} - \text{central rate} = \text{MCAs} \qquad \text{(eq. 4)}$$
$$\text{that is, } 600 - 650 = -50 \qquad \text{(eq. 5)}$$

Clearly, MCAs are negative for the country where a devaluation has taken place (and positive in Germany, a country whose currency has been revalued).

3 Implications for exports and imports

(i) France needs to levy duties on its exports to other member countries, in order to compensate for the effects of the devaluation. On the other hand, it is given import subsidies.
(ii) The revaluation of the DM makes German wheat exports dearer. Therefore, German exports need to be subsidized, and imports levied.

Covering eight commodity groups, the green rates differed according to the products involved. They have been adjustable in the long run following the realignments of parities within the European Monetary System. Introduced as a temporary measure in the late 1960s, they were to be short-lived partly because of the many distortions generated by the system on the EU agricultural market. A myriad of examples highlighting lucrative intra-EU smuggling activities have been reported. The UK had in 1986 a higher negative MCA than Ireland, as a result of the devaluation of sterling in relation to the punt. Cattle were exported from Northern Ireland to the Republic of Ireland where they were slaughtered and re-exported to mainland UK for further processing and sale. Exports of cattle to the Republic of Ireland were levied; re-exports of meat to the UK were appreciably subsidized. From the standpoint of British meat processors, this business was lucrative so long as the export levies on cattle (originating from Northern Ireland) were less than the import subsidies on meat (originating from the Republic).

The phasing out of the MCAs has been attempted on many different occasions since 1984. The only logical and long-term solution has been the irrevocable locking of exchange rates in the newly created European Monetary Union (EMU). It should be noted that they were seen as one form of 'non-tariff barrier', and that they should not have survived the removal of all remaining border controls in the ambit of the so-called 1992 Program. A more stable monetary environment in the late 1980s and early 1990s was indeed conducive to their gradual elimination. Unfortunately, the 'monetary turmoil' of September 1992, with the suspension of the Italian Lira and Pound Sterling from the EMS exchange rate mechanism, called for a temporary re-introduction of negative MCAs for a number of commodities in the case of Italy, the UK and Spain. Since that time, and until the coming into effect of stage 3 of EMU, the EU Agri-Monetary Management Committee committed itself to the dismantling of newly created MCAs by further immediate devaluation of green rates in the above countries. In September 1993, a 'switch over' mechanism was temporarily introduced, whereby all currency movements were compensated by price increases in countries with weak currencies, and price decreases in the countries with appreciating and strong currencies (a system not finding favor with German farmers).

4.1.2 Conclusion

The CAP has emerged as a compromise of conflicting views among the six founding member states. Its original principles have been market unity, community preference and common financing of its policy. Its complex instruments (target prices, threshold and intervention prices and cross-border mechanisms) apply to most of the agricultural products with a few exceptions, such as potatoes. If the phasing out of the onerous MCAs, introduced in order to ensure the principle of common price for farmers living in a Union characterized by currencies' instability, alleviated somewhat the budgetary burden, the price support policy of the CAP involved large costs. Are these costs justified by the success of the policy?

4.2 Critical appraisal of the CAP: relative success/failure of the policy

That the CAP has been successful in achieving most of its objectives, as laid down in article 39, is unquestionable. The EU has regular supplies of agricultural commodities; technical progress has taken place to a remarkable extent (see Chapter 5); markets have stabilized; earnings of economically marginal farming units would have been far less substantial had the CAP not been in place. The one objective that was not met, in light of international development, was that of 'reasonable prices to consumers', although neither did the Treaty of Rome nor did the Commission or the Council ever endeavor to define what was meant by 'reasonable prices'.

In spite of its relative success, the CAP has been decried and has been the object of repeated harsh criticisms. The market price support policy embodied in the CAP has been criticized for leading to a complicated and costly administrative system, to the distortion of resource use and allocation, to international trade frictions, and for contributing to detrimental environmental effects. Moreover, the CAP has often been condemned for favoring the most prosperous farmers in the Union at the expense of the less well-off majority. This section will appraise the drawbacks of the policy, in terms of its budgetary costs, the problem of surpluses, the costs for the EU taxpayer and consumer, and the cost endured by third nations, in particular the developing world.

4.2.1 A severe financial drain on the EU budget

Agricultural spending under the CAP is largely financed from the agricultural section of the Community budget, or European Agricultural Guarantee and Guidance Fund (EAGGF).[5] In each member state, EU financing in relation to agriculture has gradually taken over national financing. For example, in the case of Ireland, national spending on agriculture represented two-thirds of total expenditure in relation to agriculture when Ireland joined the Community in 1973; the national share dropped to 17 per cent in 1993 (Matthews, 1995). This is why the problem of costs must be considered at the European level. The EAGGF was set up by Regulation No. 25 of 1962 on the financing of the CAP (as last amended by Regulation (EEC) No. 728/70). The Guarantee section of the EAGGF covers price support policies, whereas the Guidance section is aimed at fostering structural change in the EU agricultural sector. Total expenditure under EAGGF was ECU 44,530 million in 2001, but a bare 6.8 per cent of this amount, that is less than ECU 3 billion, was spent on structural policies (CEC, 2002a). Because of the re-balancing of agricultural support towards direct payments to farmers, direct aids, to the benefit of producers, have become the major component of the EAGGF Guarantee section. During the 1999 financial year, monetary compensation to EU farmers amounted nearly to 73 per cent of total funding under EAGGF guarantee. The other 27 per cent were spent on market support policies during the same year, in the following way: export refunds (12.5 per cent), storage costs (5.2 per cent), and consumption aid (3.8 per cent), which is ironically nearly of the same magnitude as the aid to processing and marketing units (3.1 per cent).

In 2001, the total EU budget amounted to ECU 92,198 million, which still makes agriculture the biggest user of EU funds, with 48.3 per cent of the total EU budget (CEC, 2002a). This leaves a small half of the EU budget for other policies encompassed in the so-called structural funds (roughly 30 per cent of total), the cohesion funds, development and cooperation and administration. The structural funds contain the

European Social Fund (ESF), the European Regional and Development Fund (ERDF) and the Guidance section of the FEOGA. The Cohesion Funds were created in 1993 to assist the poorer member states of the EU to upgrade their infrastructure while facilitating their qualification for the EURO. They have been targeted at these countries whose per capita GNP is less than 90 per cent of the Union average, that is Ireland (at the time), Greece, Spain and Portugal. Although expenditure on agriculture has increased at a steady pace since the creation of the EEC in 1958, its share in total EU expenditures has decreased substantially (Table 4.1). In 1973, there was hardly any other Common Policy apart from the CAP, which as the pillar of European construction absorbed a healthy 80 per cent of the total. 'Deepening', that is the building of an even more integrated Europe, has given rise to a multiplication of other policies, starting with the creation of the ERDF in 1975. This process has helped reduce the dominance of the EAGGF in the EU budget. Note that expenditure on fisheries represents only slightly more than 1 per cent of EAGGF expenditure.

Since 1971, the EU budget has been financed from the EU 'own resources', which encompass custom duties, agricultural levies, and a percentage of VAT receipts.[6] In addition, levies arising from the common organization of the sugar and iso-glucose markets represent another revenue from the CAP. Finally, when the Community's own resources became exhausted in the early 1980s, a 'fourth resource' was added in 1988, known as the GNP key. Indeed, the 1982 budget ended with over spending under the EAGGF Guarantee section of ECU

Table 4.1 Evolution of EU expenditure on CAP and other policies (1973–2001, in %)

	1973	1989	1995	2001
Agriculture and fisheries	80.6	67.0	55.9	49.2
Structural and cohesion fund	5.5	16.8	23.6	33.7
Research and energy	1.6	3.4	4.6	6.5
Development/cooperation	1.4	2.3	–	4.9
Administration	5.5	5.7	5.1	4.9
Miscellaneous*	5.4	4.8	10.7*	0.8
Total	100.0	100.0	100.0	100.0

* The category 'Miscellaneous' comprises inter alia: Repayments to member states, Foreign Exchange Losses, Balance for the Previous Financial Year carried over, as well as the category 'development and cooperation' for the year 1995.
Source: European Commission, 'European Economy', Various issues.

675 million; this represented 4.3 per cent of the 'authorized' expenditure under same. Note that expenditure under EAGGF Guarantee section is generally difficult to estimate with any accuracy because of the influence of factors that lie beyond the control of the EU; these factors are the climatic variations, as well as the discrepancies in the exchange rate with the dollar.

Custom duties arise from the imposition of the common external tariff on all imports from non-member countries. Agricultural levies are variable taxes imposed on imports of agricultural commodities originating from non-member countries, and subject to a common market organization. A proportion of all VAT (value added tax) receipts represents a large financial contribution to the EC/EU budget (see Table 4.2). In line with the Council's decision of 24 June 1988, a maximum of 1.4 per cent of all VAT receipts by each member state is payable to the Community; however, this rate is applied to a VAT base limited to 55 per cent of GNP. The principle of instituting a levy on sugar and iso-glucose production means that every sugar manufacturer must pay a contribution in order to cover expenditure arising from market support. Although they are part of the resources of the CAP, they are specific to the sugar and iso-glucose markets in that they are used to maintain financial equilibrium. Their aim was to create acceptance among producers of responsibility for maintaining equilibrium (Strasser, 1992).[7] The GNP resource represents a major step forward in the financing of the EU overall activities. Representing 8.7 per cent of resources in 1989, it has been calculated by applying a specific rate to each member state's GNP, which varies from year to year. This resource is not truly a 'own resource' of the EC/EU as it is based upon each member state's ability to pay.

Table 4.2 Evolution of revenues (1971–2000, in %)

	1971	**1989**	**1995**	**2000**
Agricultural levies*	31.2	6.1	2.6	2.3
Customs duties	25.4	24.7	17.2	12.4
VAT and GNP-based resource	40.4	67.2	79.5	84.6
Miscellaneous	3.0	2.0	0.7	0.7
Total	100.0	100.0	100.0	100.0

NB: The category 'miscellaneous' comprises adjustments and surplus available, from the preceding financial year.

* Includes sugar and iso-glucose levies.

Source: CEC, 'European Economy', Various issues.

Table 4.3 shows the apportionment of EAGGF Guarantee expenditure among the different member states, in relation to the size of their national farming population. As can be seen, the Northern EU countries, with the exception of new member states such as Finland and Sweden, benefit the most from the Guarantee section of the EAGGF, and thus from agricultural spending under the EU budget. With only 1.4 per cent of the total EU farming population, Belgium received 2.5 per cent of total EAGGF guarantee commitments in 1999, yielding a ratio of 1.78. The ratio is high also for Denmark (2.46), Ireland (2.15), France (1.69) and the UK (1.62). The ratio for Germany, The Netherlands and Sweden is around 1, which suggests that the three countries score less well than their other north-European counterparts in terms of per capita EAGFF guarantee expenditure. The low ratios for Mediterranean countries (that is Portugal, Greece, Italy and Spain) are only partly explained by productivity considerations. Labor productivity reached more than

Table 4.3 Distribution of agricultural funds (EAGGF Guarantee section) by member country, compared with the relative importance of the farming community (1999, in %)

	EAGGF guarantee expenditure (1)	**Share of national farming population in total EU farming population* (2)**	**Funding per capita (1)/(2)**	**Labor productivity (in EURO)#**
Belgium	2.5	1.4	1.78	72.85
Denmark	3.2	1.3	2.46	85.80
Germany	14.5	15.0	0.96	40.24
Greece	6.5	9.7	0.67	16.86
Spain	13.2	14.8	0.89	32.69
France	23.7	14.0	1.69	65.00
Ireland	4.3	2.0	2.15	40.21
Italy	11.8	16.2	0.72	37.61
Luxembourg	0.0	0.0	–	83.66
The Netherlands	3.3	3.4	0.97	79.90
Austria	2.1	3.3	0.63	22.90
Portugal	1.7	8.9	0.19	10.32
Finland	1.4	2.1	0.66	23.87
Sweden	1.9	1.8	1.05	36.23
UK	9.9	6.1	1.62	56.95
EU-15	100.0	100.0	–	39.67

* The figures in this column encompass employment in agriculture, forestry, hunting and fisheries sectors.

This is calculated as $Q/L \times 10^3$.

Source: Calculations derived from CEC (2002a), Tables T/26 and T/105.

€37,000 in Italy in 1999; this is only marginally lower than the labor productivity ratio observable in Ireland during the same year (at €40, 200), and yet the Irish farmers receive a much larger proportion of EAGGF funding compared with their Italian counterparts. The fact that the EAGGF Guarantee section benefits mostly the north-European countries is explained by the privileged protective regime accorded to dairy products, meat and cereals, all commodities essentially produced by Northern Europe. Table 4.4 depicts the magnitude of CAP benefits accruing to a selected number of commodities in recent years.

Commodities produced in the south of the EU, wine, fruits, vegetables and olive oil receive less Community support and yield smaller gains. Consequently, per capita EAGGF guarantee expenditure is very low for the Mediterranean countries, who tend to have a relatively large fruits, vegetables and wine sectors. A clear correspondence exists thus between Tables 4.3 and 4.4.

The poor position of Italy, Portugal, Spain and Greece in terms of EAGGF guarantee payments is slightly compensated by more favorable payments under the Guidance section of the Fund. In 1999, Spain received 27.4 per cent of total funds allocated through this section; Greece received 10.4 per cent, Portugal 9.8 per cent and Italy 9 per cent. However, it is interesting to note that the shares of the Guidance section going to both France (16.6 per cent of the total) and Germany (14 per cent) were higher than those received by most of the Mediterranean countries during that year (CEC, 2002a).

Although it is possible to talk about a broad north–south divide in the case of European agriculture, huge disparities exist within the various

Table 4.4 EAGGF guarantee expenditure by product (2000, in %)

Products	
Cereals	40.59
Beef, veal, sheep and goat meat	15.68
Milk products	6.67
Olive oil	5.34
Fruits and vegetables	4.03
Wine	1.70
Sugar	4.87
Other products	21.12
Total	100.00

Source: CEC (2002a), Tables T/106 and T/107.

countries. Some 11.2 per cent of farmers in the top two size classes (that is above 40 ESU, European Size Unit) capture nearly 44 per cent of all CAP benefits (European Commission, 1994b).

The price support policy of the CAP, as encapsulated in the EAGGF Guarantee section of the EU budget, still represents half of all EU expenditures. This share used to be much higher in the past. CAP spending has essentially benefited the producers of cereals, dairy products and meat located in northern Europe, at the expense of the southern producers. If spending on agriculture has been monitored and curbed in the last 20 years, more efforts need to be made. In particular, a very marginal proportion of the EU budget (that is less than 10 per cent) goes to the Guidance section. This share would need to be raised substantially if agricultural holdings in the EU stand a chance to compete on the international liberalized market. However, all the figures given above should be replaced in a broader perspective. Overall, the general budget of the EU is minuscule. The EU budget represents no more than 1.27 per cent of total EU GDP, and approximately 3 per cent of the aggregate EU budgets. A change in the structure of the budget with the introduction of the GNP key, and with the doubling of structural funds emerged in the late 1980s. New movements in that direction need to be undertaken.

4.2.2 Costs in terms of high prices and surpluses

The generous support policy led to large surpluses, the disposal of which has been extremely costly. There seems to be a universal agreement on two propositions concerning the CAP of the 1970s and 1980s (Duchêne *et al.*, 1985):

> Proposition 1: The CAP has been suffering from a chronic condition of oversupply.
> Proposition 2: This has been due to high farm prices.

Proposition 1 is easy to demonstrate. The criteria used to that purpose is the self-sufficiency rate: it is defined as the production to consumption ratio. Any product with a ratio above the critical value of 1 (or 100) is in surplus. Self-sufficiency rates for a number of agricultural commodities are given in Table 4.5. Oversupply of milk products seems indeed to be chronic. As can generally be seen in this table, the situation has not improved dramatically since the mid-1980s. On the contrary, self-sufficiency ratios were substantially higher in the early 1990s for a number of commodities such as wheat (and cereals in general), sugar, wine and butter. Ratios are below 100 for goat and sheep meat, fresh

Table 4.5 Self-sufficiency rates in certain agricultural products (1985/86, 1992/93 and 1998/99)

	1985/86	1992/93	1998/99#
Wheat	124	133	120
Barley	119	123	124
Sugar	123	128	–
Fresh vegetables	107	106	–
Wine	104	115*	109
Whole milk powder	316	272	201
Cheese	106	107*	105
Butter	110	121*	99
Total meat	102	102*	107
Sheep/goat meat	80	81*	83

* Figure for 1989/90 and # EU-15.
Source: CEC (2002a), Table T/204.

fruits, maize, and vegetable oils and fats. The late 1990s are characterized by an increase in self-sufficiency ratios for meat products in general.

Proposition 2: surpluses are due to high farm prices.
The reasoning is as follows: technically, there is a shift of the supply curve towards the right (as a result of technological change), combined with a low income elasticity of demand for food (see Chapter 2). This necessitates judicious adjustments, such as leaving the price perform its market clearing function. In the case of the CAP, surpluses have developed because prices are determined by political decisions rather than by market forces.

It is first of all necessary to specify what type of price one has in mind, and it is necessary to make a distinction between the following:

- real prices versus nominal prices,
- entry prices versus support prices,
- when comparing the EU and the USA, for example, to take into account the distortions brought about by exchange rates.

- *Real versus nominal*

The real price is the nominal price adjusted for inflation. The EU Commission produced evidence that between 1970 and 1985, farm prices in real terms fell continuously in all but one of the founder states of the Community. Prices rose indeed in Italy in the early seventies, as the country was catching up with its EU counterparts. A similar phenomenon has been observed in the case of poorer member states during the transition period; this was for example the case of Ireland up to 1978. Table 4.6 gives an insight into the evolution of real prices in the EU since 1985. As can be observed, statistical evidence suggests that real prices in general have declined by 25 per cent between 1985 and 1993, and by a further 25 per cent over the 1990s. The decrease has been less important for wine, fresh vegetables and fruits.

Entry versus support

The support or intervention prices are normally lower than the entry prices, which have a protective purpose. However, it is undeniable that support prices, although less than entry prices, have been for a long time well above world prices.

Distortions due to exchange rates

American grain and meat prices are lower than European ones, but whether the prices of dairy products and cereals in the Community or the USA are higher or lower, depends also on the €/$ exchange rates. A deteriorating dollar would make EU products even dearer, and would nourish trade tensions.

Real prices in the EU have declined continuously since the mechanisms of the CAP have been fully in operation. In spite of their decline, they tended to remain generally above the prices prevailing in other western

Table 4.6 Evolution of producer prices for agricultural products in the EU (1985–93 and 1990–98, VAT excluded)

	Real index: 1985 = 100	**Real index: 1990 = 100**
Cereals and rice	65.5	58.4
Sugar beet	79.8	85.3
Fresh vegetables	87.2	89.9
Fruits	71.8	87.7
Wine	86.0	95.3
Seeds	87.2	83.7
Beef	78.5	na
Milk	85.3	81.2
Total	75.0	76.5

Source: CEC (1995, 2002a) The Agricultural Situation in the European Union. Derived and adapted from Tables T/69 and T/87.

economies. This is due to a sophisticated protectionist policy partly aimed at insulating the EU agricultural market from that of the rest of the world. Being sheltered from external influences, the EU farmers could consolidate their position and remain in business, although only a fraction of them were able to enjoy high incomes. Although it is possible to infer that prices in the EU are relatively high (in spite of their decline), it is presumptuous and erroneous to assert that prices alone are responsible for surpluses developing in the EU agricultural sector. A balancing item is demand. Where demand is static or declining, even a fall in production, as a result of declining prices, may produce surpluses. Duchêne *et al.* (1985) use the example of Italy after EC membership to show that higher prices were concomitant with declining self-sufficiency ratios. The production to consumption ratio stood at 91 per cent in 1959, and declined to a low 78 per cent in 1973. In the meantime, producers' prices rose by a quarter. The explanation was that although production increased by a third, consumption increased even faster, by half during the same period of time. A high income elasticity of demand prevented surpluses from building up.

4.2.3 Costs for EU consumers and taxpayers

The extravagant costs borne by both consumers and taxpayers in relation to the benefits received by producers have been discussed in length (Roningen and Dixit, 1989; OECD, 1992), and Chapter 8 will provide a more thorough account of this problem. Balassa's own words summarize the situation before the beginning of the Uruguay Round:[8]

> It has been estimated that in 1985 the cost of agricultural policies to the consumer and tax payer was 18 per cent higher than the benefits the producers derived from protection in the European Common Market, while it was 221 per cent in Japan and 6 per cent lower in the United States. (Balassa, 1988: 162)

The best-known measures of agricultural policy support are the nominal protection coefficient (NPC), and the producer and consumer subsidy equivalents (PSEs and CSEs respectively). The NPC is simply the ratio of the price received by the domestic producer to the world price of the same product, that is:

$$NPC = \frac{P_{\text{prod}}}{P_{\text{world}}}$$

If $NPC > 1$, this means that domestic producers are protected by their national government.

PSEs and CSEs are a common measure of agricultural policy impact on purchased input prices or output prices, and, by extent on agricultural incomes. They are termed 'subsidy equivalent' because, although they are not all subsidies, they have a price effect which is comparable to that of subsidies.[9] In particular, the PSE indicator gives an indication of the monetary transfer from consumers and taxpayers to agricultural producers, as a result of agricultural policies. The concepts of PSE and CSE are logically bounded in a zero-sum game framework. They imply that whatever is earned by a producer, in excess of what is possible under free trade, is lost by both consumers and taxpayers, and vice versa. The principle underlying these measures is that the benefits arising from protection are pocketed by a category of economic agents, but have to be disbursed by other groups of agents in the economy. The OECD (Organization for Economic Cooperation and Development) defines PSEs and CSEs as follows:[10]

> the PSE measures the value of the monetary transfers to producers from consumers of agricultural products and from taxpayers, while the CSE measures the value of monetary transfers from domestic consumers to producers and taxpayers (OECD, 1992: 231–2).

The PSE measure quantifies the influence of 'direct' farm policies, although it leaves out some 'indirect' measures. For example, in the case of the EU, are not included in this indicator environmental payments, welfare expenditures, and outlays for stockholding (European Commission, 1994a). The direct implication of price support policies is to raise consumer prices. The CSE is an indication of the monetary transfer from domestic consumers to producers and taxpayers, as a result of a given set of agricultural policies (Cahill and Legg, 1989–90). Although very close to the previous indicator, in the case where domestic farm gate prices are used, the CSE measurement differs from the PSE index in that it is limited to the effect of agricultural policies on consumers only. In 1992, the total transfers from consumers and taxpayers arising from the implementation of the EU agricultural policy were as depicted in Table 4.7.

In Table 4.7, transfers from taxpayers include all budgetary outlays that are encompassed in the PSEs as defined above. These transfers were more than ECU 51 billion in 1992, but when budget revenues, as defined strictly by duties on agricultural imports were taken into account, these transfers were negligible. Much more substantial were the transfers from EU consumers, transfers that correspond to the total

Table 4.7 Total EU agricultural policy transfers (ECU billion)

	1988	1992	OECD average
From taxpayers	38.7	51.8	
(less EU budget revenues)	−0.9	−0.6	
From consumers	64.6	69.3	
Total	102.4	120.5	
(share in total GDP)		2.0%	2.1%

NB: Budget revenues refer to tariffs on agricultural imports.
Source: European Commission, 1994a, p. 90.

market price support element of the CSE, net of consumer subsidies borne by taxpayers (European Commission, 1994a). Amounting to more than ECU 69 billions in 1992, these transfers can be viewed as colossal since they represented more than 1 per cent of total EU GDP in 1992. Total transfers from EU taxpayers and consumers to EU producers represented 2 per cent of GDP in the early 1990s, a figure that compares well with that of the countries belonging to the OECD area (2.1 per cent in the same year). However, it should be borne in mind that the extent of protection varies to a very large degree in the OECD area, with Canada and non-EU Western Europe having a net benefit to cost ratio of 16 and 59 per cent respectively (Roningen and Dixit, 1989).

4.2.4 Costs for developing countries

Since protection of the EU agricultural sector acts as a deterrent to imports from third countries and tends to depress world prices, the benefits accruing to the EU producers are earned at the expense of producers in other nations. These other nations are the developing countries where the protection of the 'infant' manufacturing sector has normally discriminated against the agricultural sector (World Bank, 1986), and where farmers are normally burdened with taxes and poor terms of trade. The EU gross producer benefits that offset policies of other nations has been estimated at \$12.5 billions for the year 1986 (Roningen and Dixit, 1989). This amount was slightly above that of the USA (\$11.3 billion). Broadly, the cost of support to the agricultural industry of the OECD countries was 250 per cent higher than the same countries' aid to the developing countries in the 1980s.

4.2.5 Conclusion: a cost/benefit analysis?

In general, the CAP led to negative achievements, such as high prices borne by consumers and taxpayers, surpluses, misallocation of resources, heavy burden on the EU budget, costs to other nations and detrimental effects on the environment. However, the positive results of the CAP should not be understated. The EU has been able to reduce its dependency from other nations of the world; the objective of common price has been held; and incomes have been sustained, albeit at a lower level than those received in other sectors of the economy. Probably one of the most disturbing inefficiencies of the policy has been its inability to cope with intra-EU regional disparities. To some extent, the CAP has intensified regional imbalances by favoring the dairy, cereals and meat (pig and poultry excepted) sectors. The EAGGF Guidance section is far too small to compensate for these disparities. The problems of chronic oversupply and high costs, as well as pressures building up on the international arena, have led to a step-by-step reform of the CAP.

4.3 CAP reforms and the future of the policy

Stimulated by a generous price support system as well as by technical change, and constrained by a low income elasticity of demand, EC farm output increased by 30 per cent between 1963 and 1973 (European Commission, 1994b). Self-sufficiency was attained in most products with the exception of the income sensitive commodities such as beef, vegetables and fruits. In less than 25 years food shortages turned into surpluses, as demand failed to keep pace with an abundant supply. The gigantic production surpluses resulted in rapidly growing Community expenditure on intervention purchases and export subsidies for dairy products, wheat and sugar. It became clear that the policy of market control could no longer be based on guaranteed unlimited sales at fixed prices for practically all products. If the late 1960s and 1970s are the years of embryonic changes, the 1980s and early 1990s are characterized by a definite progress towards a more rational ordering of the European and international agricultural sectors. The Uruguay Round of GATT negotiations (1986–93) has certainly exerted an influence on this reordering, as we will see in Chapter 8.

4.3.1 A whispering reform: 1968 to 1983

In order to cope with the problem of oversupply at the European level, proposals started to be made in the late 1960s and early 1970s. The Mansholt Plan, named after Sicco Mansholt the then EU Commissioner for Agriculture, is regarded as a precursor of policy reform, although it had little effectiveness. The Mansholt plan attempted to use prices more flexibly to achieve a better balance between demand and supply. Under the plan, 5 million hectares of land and 5 million people would have left the EC-6 agricultural production during the 1970s. It also sought to convert dairy herds into beef production, which at that time was much lower than consumption (European Commission, 1994a: 13). The plan proved too ambitious, and too radical to be politically and socially acceptable. Had it been put in operation, it would have implied the destruction of many family farms. However, the plan gave birth to three socio-structural directives adopted in 1972. They dealt respectively with subsidies for farm modernization, cessation of farming activities and land reallocation, as well as with the supply of training facilities aimed at skills improvement. The directive on farm modernization proved very popular among the better-off farmers living in the most prosperous areas of the Community.

Probably the major positive implication of the aborted Mansholt plan was that it established the principle of CAP reform in Community policy thinking. Although the beginning of international negotiations under the Uruguay Round in the second part of the 1980s increased the speed of agricultural reform in the Community, other endogenous factors demanded a major review of the CAP, in particular:

(i) the deepening of the EU budgetary crisis in the years following the first enlargement,
(ii) an unfavorable macroeconomic climate,
(iii) the attitude of the UK.

(i) CAP spending doubled in real terms, and increased four-fold in nominal terms, between the mid-1970s and the mid-1980s (Figure 4.2). The EAGGF guarantee expenditures doubled again in nominal terms between 1985 and 1995. Budgetary costs increased rapidly during this period of time, due to intervention purchases of surpluses and to subsidized exports.

As stated above, growing surpluses in all main crop and livestock sectors went in hand with repeated currency instability. The resulting rising MCA expenditure placed an additional pressure on costs. As per capita consumption had reached satisfactory levels in Europe, it was expected that the enlargement of the EU to incorporate the UK would make it easier for EU farmers to dispose of surpluses. Soon after the accession of the three new member states, self-sufficiency in a broad range of products (wine, cereals, dairy products and meat) indeed declined. However, the first enlargement was only a very short-term palliative for the CAP growing surpluses, as they increased again at the end

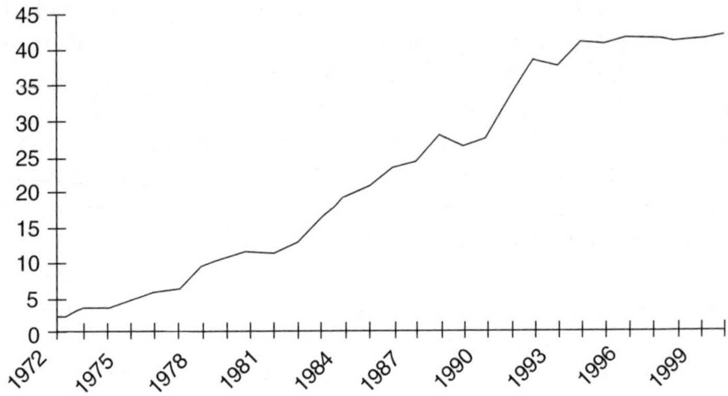

Figure 4.2 EAGGF Guarantee expenditure (1972–2000) (Bn ECU/EURO, nominal prices)
Source: CEC (various issues) European Economy.

of the decade. The only novelty was that the difficulties of the six were now shared among nine member states.

(ii) The Arab–Israeli war provoked an escalation of oil prices: between 1972 and 1974, the price of the barrel of oil increased from $2 to $11. In Europe, production costs were rampant, inflation rates were soaring, monetary instability was aggravated, unemployment rose, and the growth of national income decelerated. In particular, rising unemployment rates jeopardized the adjustment process in the European economies, as it made it more difficult for people leaving the farm to find jobs elsewhere. At the same time, agricultural prices were on a continuing decreasing trend. This necessitated a further increase in nominal protection, although it was difficult to tighten intervention rules. Squeezed between stagnant prices and increasing costs, the farming population's discontent increased.

(iii) Since the UK was a net importer of foodstuffs, and since the EAGGF represented at the time roughly 80 per cent of the Community budget, it was clear that the UK would not be a beneficiary of European integration in budgetary terms, and would call for lower price support and alternative common policies.

In this background, the first 'market-oriented' proposals were made (van Riemsdijk, 1972; Koester and Tangermann, 1976; Tarditi, 1984). These proposals reserved the instrument of price policy for clearing the markets, maintaining some form of market regulation to stabilize prices.

The Commission embarked then upon a series of reports some of which are synoptically presented in Table 4.8. Note that the euphemisms 'improvement' and 'guidelines' are preferred to the more radical word 'reform'. The 1973 document explicitly announced that the price level considered by the Commission would be based on costs and returns of modern farms ('objective criteria' principle). In the context of increasing inflation rates, and consequently of high price requirement, the new policy would have been subject to a 'number of uncertainties'. These proposals did not prove satisfactory. More fruitful were the other two documents. The 1980 document widened the application of the co-responsibility levy, whereas the 1981 document refined it further. The principle and practice of applying a levy on sugar and iso-glucose had been in existence since the inception of the CAP, whereas the same principle was introduced in the milk sector in 1977. What is new in the 1980 and 1981 documents are the wording 'co-responsibility', as well as the ambition to make this mechanism a tool *à part entière* of agricultural policy, with a wider application. A co-responsibility levy implies that producers are responsible for over-production and should therefore bear the cost of the surplus. In line with the recommendations made in the 1981 Commission document, the principle of 'guarantee thresholds' for a range of products was introduced. This entailed full CAP benefits up to a pre-specified level of output. Failure to comply with these pre-set levels of output would have implied either one of the following:

(i) a reduction in target and intervention prices,
(ii) a reduction in the aid paid under market regulation,
(iii) an imposition of a 'co-responsibility' levy,
(iv) the imposition of production quotas at the national or enterprise level.

Table 4.8 Early Commission's reports on the review of the CAP

Date	Title	Major principle	Results
1973	Improvement of the CAP	'Objective criteria principle'	Little application Not satisfactory
1980 – December	Reflections on the CAP	Extends the principle of 'producer co-responsibility'	Wider application
1981 – October	Guidelines for European agriculture	Guarantee thresholds	Wider application

The Commission's price proposals for the marketing year 1982/83 were based upon the co-responsibility principle.[11] Despite the many difficulties in gaining unanimous support, the principle of a co-responsibility levy was introduced in the cereals sector much later, in 1986. It appears that in the midst of a deep budgetary crisis, the Community responded with fringe/marginal corrective measures summarized as a smaller rise in nominal price support (which corresponded to substantial decreases in real terms), and a tightening of intervention rules. A more comprehensive and assertive approach, stating more thoroughly the new orientations was embodied in the SIENA Memorandum (1984).

4.3.2 1984 and onwards: a more decisive move towards market reordering

Whereas the budgetary crisis was threatening the CAP with quasi-imminent financial collapse in 1984, a more decisive stance was gradually taken at various Council meetings, which culminated with the Siena Memorandum of March 1984. The recommendations of the Memorandum were:

(i) a more decisive move towards market-oriented policies,
(ii) an income compensation for farmers,
(iii) the phasing out of the MCAs.

(i) Market-oriented policies

The move towards market-oriented policies entails less protection and fair prices. The EU chose three lines of action aimed at reducing the level of price support.

- an average price cut of 0.5 per cent applied during the 1984/85 marketing year,
- the extension of the 'guarantee threshold' to the cases of durum wheat, sunflower seeds and raisins, and
- the introduction of a quota-system in the milk sector.

The 'guarantee thresholds' are also referred to as 'stabilizers'. This system implies an automatic reduction in the intervention price in case the production exceeds a certain ceiling. Production quotas involve fixing maximum production levels for each country, and within each country setting a maximum authorized production level for every farmer or cooperative. Limiting production in this way prevents prices sinking, and reduces intervention costs. Introduced in the liquid milk sector in

April 1984, they had an immediate impact on milk production, and on dairy cow numbers in the EU. Prior to 1984, production quotas had already been applied to sugar beet, and they were subsequently extended to other products such as cereals, oilseeds and processed tomatoes. In the long run, production quotas are seen as a means to reduce prices.

The price support mechanisms had ensured a minimum income for EU farmers in the past. Its fundamental revision was deemed to create income problems for farmers, in particular for those living in the less favored areas of the Community. This called for the inclusion of an income-compensation policy.

(ii) Income compensation for farmers

Various schemes were to compensate farmers for the income loss incurred as the result of the decline in price support. This compensation is worked out on a regional basis according to local, economic and social conditions.

(iii) The phasing out of the MCAs

The need to eliminate the many distortions created by the MCAs was reiterated. The method used to reduce MCAs was to devalue the green rate. The phasing out of negative MCAs led to an increase in the intervention price, whereas positive MCAs, that is those associated with strong currencies, were hard to dismantle, since this involved a farm price cut in national currency terms. This was an unpopular result for the farming community of wealthier countries in the background of the crisis affecting farm incomes. In 1984, the Council undertook to ensure that any change in monetary parities would not give rise to new positive MCAs for strong currencies (European Parliament, 1994). Consequently the MCAs were no longer calculated as the gap between the green rate and the central rate, but instead the gap was adjusted by a 'correcting factor' known as the 'green central rate'. This 'correcting factor' is a numerical constant that expresses the percentage of appreciation of the strongest currency, when parity realignments take place. Figure 4.3 illustrates the new method of calculation in the case of France (where MCAs are negative) and Germany (positive MCAs). According to the system, new positive MCAs have been avoided by means of an increase in the level of common prices at the time of a monetary realignment. This increase is proportional to the highest upward revaluation against the ECU (normally the DM).

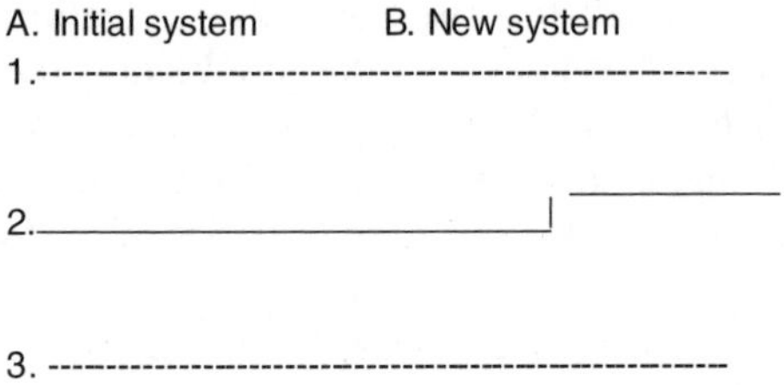

1. Price in German Marks determined according to the green rate.
2. Community prices in ECU, determined according to the:
   (i) central rate, under the initial system A,
   (ii) green central rate, under the new method B.
3. Price in French Francs, determined according to the green rate.

Figure 4.3 New method of calculating MCAs since the 1984–85 marketing year
Source: European Parliament (1994: 410).

This new system produced a 'green ECU' which corresponded to the real ECU (or EMS ECU) adjusted by the 'correcting factor'. As a result of this system, the nominal value of a price in ECU remained unchanged, but its equivalent in national currency increased, except for the one that underwent the highest revaluation.[12] A certain stability of Community currencies expressed through less frequent currency realignments in the late 1980s and early 1990s allowed this system to be successful. Whereas positive MCAs have been rendered technically inoperative since 1984, most of the remaining negative MCAs were gradually removed at the end of the 1980s. In July 1991, MCAs had been removed in the member countries with almost stable currencies (Germany, Benelux countries, France, Denmark) as well as in Italy, Ireland, the UK and Portugal (Baudin, 1991). Only Spain and Greece kept some green rates for a limited number of products (cereals, beef in particular). As a result of the realignments made on 30 January 1993,[13] the 'correcting factor' rose to the value of 1.205454 (that is an increase of 5.2 per cent, compared with the pre-realignment situation). Technically, the green ECU was revalued upwards by 20.54 per cent against the real ECU. This value must cancel out in the short term, entailing a price cut in ECU terms. Also, new negative MCAs were reintroduced temporarily in the case of the UK and Italy. The widening of the margins of fluctuations around central rates to ±15 per cent since August 1993, has rendered realignments more unlikely. The SIENA Memorandum paved the way for a more decisive change in the philosophy and direction of the CAP, which culminated with the 1992 Reform.

4.3.3 The 1992 CAP reform: a more fundamental change

In spite of a slow but gradual progress, which did lead to adjustments in some markets, CAP spending was up by a further 30 per cent in nominal terms between 1989 and 1991. Net exports of cereals had grown enormously, whereas 'beef mountains' started to pile up. External pressures for a more substantial reform were increasingly felt in the context of the international Uruguay Round negotiations in the GATT (see Chapter 8). In February 1991, the Commission produced some *reflections* on the future of the CAP (CEC, 1991a). These were followed by a set of *proposals* in July of the same year (CEC, 1991b), which finally gave rise to subsequent *recommendations* that the Council of Ministers adopted on 22 May 1992 (OJEC, 1992). These proposals and recommendations became widely known as the 'Mac Sharry' package of the reform, named after the EU Commissioner for Agriculture at the time. It should be noted that various iterations have led to the Mac Sharry proposals; these iterations will be seen in greater detail in Chapter 8.

Box 2 The major elements of the 1992 CAP Reform

- Cuts in agricultural support prices and 'tariffication':

  (i) a 15 per cent reduction in the intervention of beef from July 1993,
  (ii) a reduction of roughly one-third in the cereal intervention price,
  (iii) cuts in the price of milk products, fruit and vegetables, sugar, tobacco and wine,
  (iv) elimination of price support for oilseeds and protein crops, and
  (v) all import levies and quotas must be transformed into *ad valorem* tariffs.

- Set aside policies applicable to producers of arable crops (cereals, oil seeds, soybeans, sunflower seeds) as well as to producers of protein crops (peas and field beans).
- Compensation through direct payments to farmers.

Box 2 (Continued)

- Re-allocation of the labor resource through early retirement schemes.
- Alternative rural development strategies:

  (i) agri-environment package,
  (ii) afforestation of agricultural land,
  (iii) 'special' member states programs, with 75 per cent of cost borne by the EAGGF Guidance section of the EU budget in Objective 1 regions.

The 1992 CAP reform represents a major change in the type of support: it definitely shifts the emphasis from price support to income compensation. Also, these proposals are contained in a global approach since, for the first time, environmental and qualitative elements are taken into account and are greatly stressed. It finally suggests a long-term perspective for the future of the CAP. Most of the changes involve a transition in three steps over the 1993–96 period of time. It should be noted however that all in all, the Mac Sharry package applies only to 50 per cent of total EU agricultural output (European Commission, 1994a). The major components of the reform are contained in Box 2.

The move towards a market-led approach, and the necessity to implement a strict management of agricultural markets were reinforced. The 30 per cent decrease in the cereal intervention price should be seen as a broad guideline. For example, the intervention price for cereals was fixed as follows:

ECU 117 per ton for the 1993/94 marketing year,
ECU 108 per ton for the 1994/95 marketing year, and
ECU 100 per ton for the 1995/96 marketing year.

The price of ECU 100 was calculated as the anticipated stabilized world market price.

The transformation of levies and quotas into *ad valorem* tariffs is aimed at restoring world market links. *Ad valorem* tariffs are more transparent than quotas and variable levies; a further step is to reduce these new tariffs gradually. Producers of arable and protein crops have the option to participate in the set-aside scheme. Applied from the 1993/94 marketing year onwards, it affects 15 per cent of the land. The land set aside shall normally be subject to rotation. It is expected that the trend of support prices towards world market prices will remove, in the long term, the necessity to have quantitative restrictions (such as milk quotas) as well as set-aside policies. In return for lower support prices and for introducing set-aside policies, farmers get compensatory payments. The magnitude of the compensatory payment is fixed per hectare, depending upon the region. Consequently, the reordering of the EU agricultural markets over the past decade and the rise in compensatory amounts explain the shift of EAGGF guarantee expenditure away from market support, as was stipulated above.

Producers of crops are compensated through direct 'area payments' calculated on the grounds of historical base areas and regional yields (European Commission, 1994a). Excluded from this scheme (and obviously from the set-aside requirement) are the smaller farmers producing less than 92 tons of cereals per annum. It is interesting to note that a small producer is defined on the basis of the output produced (number of tons) rather than on the basis of the area used. This is in turn worked out on a regional basis: an average level of output is calculated for each region of the Union. Any farmer whose level of output is below the regional level is a 'disadvantaged' farmer, and would therefore qualify for compensation. Producers of livestock products are compensated through direct 'headage payments' or premiums. These are subject to individual limits per holding as well as to regional reference herd sizes.

A gradual characteristic of direct payments to farmers is that they need to be completely 'decoupled' from past levels of production and use of production factors. The direct payments have met strong resistance among the European farming community as they have often been criticized for being the equivalent of the 'green dole'. Initially, compensatory payments were financed by the EU budget, but after seven to ten years, it is expected that member states would take over this initiative.

Early retirement schemes are supposed to promote the idea of retirement to older farmers who would be willing to pass their agricultural unit on to younger generations. They are partly aimed at improving agricultural structures through the cessation of land to more technologically driven and larger farmers. The basic yearly pension proposed was ECU 4000. This very modest level, coupled with both the philosophy of 'farming as-a-way-of-life' in some regions, and with the high opportunity cost of selling the land in other cases, raises many questions as to the probable success of the scheme (European Commission, 1994b). According to the Mac Sharry package, the action on prices is still thought to be crucial, but this action is now conceived in a wider strategy comprising an environmental approach. The speed of the CAP reform in the early 1990s has necessitated the strengthening of the so-called 'accompanying measures'. Already envisaged by the European Council of February 1988, these measures are designed to ensure that economic and social cohesion is reinforced through fully safeguarding the position of the vast majority of farmers in the EU. The Commission's 1991 document states the fact that these measures had in the past only a marginal application, and advocates the need to include them as an integral part of the new market organizations (CEC, 1991b: 8). The document also calls for the need to ensure 'an environmentally sustainable form of agricultural production and food quality' (CEC, 1991b: 11). It is interesting to note that the new philosophy embodied in the 1991 document, and taken on board slightly later by the Council as part of the 1992 Reform, is promoted by the Commission while it reasserts the principles of the CAP: single market, community preference and financial solidarity.

The move towards new economic activities in the countryside in order to reduce negative externalities and to improve land conservation and balanced development, or/and the shift towards product differentiation are the main thrust of these 'accompanying measures'. Rural development had been on the agenda of the European Commission for a number of years (CEC, 1988). In its 1988 document, the Commission considered three principles that should be integrated in any policy on rural development in Europe:

(i) social and economic cohesion as a response to wide regional disparities,
(ii) environmental conservation and maintenance of the Community's natural heritage,
(iii) adjustment of European agriculture to international markets.

These principles have been taken on board since they have led to: (i) the reform of the EC/EU budget in 1989 with the doubling of the Structural Funds, (ii) the 1992 CAP Reform with the integration of a social dimension (direct compensation to farmers), and of an environmental component.

The role of the farming community in the protection of the environment has been fully integrated into the 1992 agri-environmental program (Regulation No. 2078/92, of 30 June). Financial incentives exist that are designed to promote the use of less intensive, less polluting and more environmentally friendly production methods; they are also aimed at encouraging farmers to re-establish the diversity and quality of the natural ecosystem through appropriate management of farmland. Training of the farming community for the purpose of protecting the countryside is also a major aspect of this program (European Commission, 1994b). Governments have echoed the new European strategy in launching nation-wide plans for the promotion of rural development. For example, the need to draw the main lines of what could be seen as *Integrated Rural Development* in the case of the Irish economy, had transpired already in the second part of the 1980s (Conway and O'Hara, 1985; O'Malley, 1992). Some studies have confirmed that agri-tourism has increasingly and successfully become an alternative to farming activities in some regions of the EU (Marinelli *et al.*, 1990). Agri-tourism has been successful in creating employment outside the farm, but on the land, with positive impacts on rural incomes.

Farm diversification, or the movement away from surplus products towards income elastic commodities had already taken place in the late 1980s. From 1987/88 to 1991/92, an aid per hectare was introduced for indica varieties of rice in order to encourage the shift to income sensitive types of rice (CEC, 1993). A recent EU program (the Operational Programme for Agricuture, Forestry and Rural Development 1994–99) provided grant-aided investment for a wide range of alternative enterprises, rural tourism and horticultural enterprises. Alternative enterprising activities such as horticultural products, horses, deer, rabbits, ostriches, goats, as well as organic farming, were given assistance. For example, the Irish government has provided funds for the establishment in 1990 of the Organic Farming Unit which provides research, training and consultancy services. Moreover, the afforestation of agricultural land is also a possible way of increasing differentiation in farms. However, the major difficulty connected with afforestation of farmland is the long time span of the production cycle before any return can be envisaged. Afforestation could possibly reduce the timber deficit of the EU (European Commission, 1994b). Finally small craft industries have been encouraged. These were already encompassed in the Leader program, which was aimed at encouraging groups in local communities to organize themselves and to draw up plans for development in their region. New activities include training, rural tourism and marketing of local produce.

All these alternative rural development strategies are made possible through the structural and cohesion funds, the aim of which is to allow assistance and aid those most in need from a social and economic point of view.

4.3.4 The aftermath of the 1992 reform and the BSE (bovine spongiform encephalopathy)

The first encouraging results of the 1992 reform could be perceived almost immediately. Various sources brought evidence that surpluses have plummeted in the EU for the first time ever. In September 1995, stocks of skimmed milk powder were non-existent, except in Ireland, and stocks of butter had all nearly disappeared with the exception of the UK, Spain and Ireland. Stocks of beef started shrinking in 1993 (Le Monde, 1995) and in the mid-1990s intervention stocks were near zero (CEC, 2002b). Unfortunately, the efforts to redress the disequilibrium in the EU beef market, and the parallel depletion of beef stocks, were suddenly halted in 1996 by an unprecedented 'beef crisis' in Europe. During that year, an animal disease called *bovine spongiform encephalopathy* (BSE), and commonly known as 'mad-cow disease', a disease that had been around for some time, was for the first time acknowledged to be associated with a human fatal illness, the Creutzfeldt-Jakob Disease (CJD). Identified first in Britain in 1986, BSE was primarily caused by the use of commercial cattle feed concentrates containing meat and bone meal derived from sheep presumed to have been infected with scrapie, which is a fatal and degenerative disease affecting the central nervous system of sheep and goats. The UK beef industry was the most severely hit industry in the EU, with some 170,000 cases of BSE reported in a herd of 12 million up to 1996. However, given the principle of free movement of agricultural inputs and products within the EU market, the contagion of contaminated feed affected most EU economies, and what was a pure British problem soon became an EU wide crisis with increasing numbers of BSE cases recorded in various EU countries. The BSE outbreak may have gone almost unnoticed, had it not been for the scientific evidence brought to light in 1996 and proving that the disease could be transmitted to humans, taking the form of the CJD. As a result, consumption of beef plummeted in the UK, import controls were reinstalled within the EU market, and a worldwide ban on British beef was introduced in March 1996. In 1996, the domestic consumption of beef in the UK declined by almost 20 per cent compared with the previous year. The effects of the ban on British beef spilled over to other EU supplying countries, such as Ireland in particular. This was explained by the inability, or perhaps unwillingness and extreme caution, of many consumers from Eastern Europe and the Middle East to distinguish between British and Irish products.

In order to surmount the crisis and to restore consumer confidence, the British Government undertook a massive BSE eradication plan, aimed at removing the infected cattle from the market and at curbing beef supplies. Proposals at EU level included the necessity of introducing a draft legislation on the strict and clear identification of beef products in the EU. The labeling of EU beef products would allow the unequivocal identification of the country of origin of the product, guaranteeing thus the quality of the meat on sale. This would result in the building up of detailed computerized data bases, equivalent to an animal passport, allowing information on the animal, fattening methods and feed history. Chapter 5 will show how the EU idea of labeling and 'traceability' of food products has gained ground in recent years. Finally, and as in any crisis situation, a financial compensation package has been made available to the EU farmers affected by the BSE crisis, so as to help them recover their losses.

The beef crisis aggravated the problem of surpluses in the EU. In September 1996, it was estimated that EU farmers were rearing some 2 million surplus cattle (European Report, 1996). Although beef consumption in the EU had stabilized at a level 10 per cent lower than before the beef crisis, the EU agriculture ministers were urged to find extra financing for buying in up to 700,000 tons of beef in 1996, reintroducing thus intervention in this market. Intervention buying limits increased from 400,000 tons to 720,000 tons between 1996 and 1997. The Beef Reform Proposals suggested by the EU Commissioner for Agriculture Franz Fischler in December 1996, were to cut beef production without any detrimental implications for the EU budget. The increase in beef stocks and the intervention costs associated with this policy, had to be compensated by a corresponding decrease in other support and intervention policies. Consequently, the EU cereals and oilseeds budgets were to be reduced accordingly. An equilibrium has lately been restored on the beef market, as the stocks taken into intervention during the BSE outbreak are disappearing even faster than was anticipated initially (CEC, 2002b).

4.3.5 Agenda 2000 and the CAP as an evolving policy

The 1999 European Council in Berlin agreed the Agenda 2000 reform of the CAP. This represents an extension of the 1992 reform, favoring market-oriented policies, and consolidating rural development as a second pillar of the CAP (CEC, 2002b). This further reform is necessitated by the prospect of enlargement to Central and East European countries with large and inefficient agricultural sectors, that would, in the absence of further reform, require an unsustainable level of expenditure (see Chapter 6). The objectives of Agenda 2000 are in line with the Sustainable Development Strategy agreed by the European Council at Göteborg in 2001. These economic, social and environmental objectives stress the importance of creating a 'European Model of Agriculture' that preserves the diversity of farming systems throughout Europe, that allows increased competitiveness, food safety, stabilization of agricultural incomes and environmentally friendly production methods. Agenda 2000 calls for further price reductions for cereals and beef, compensated by increased direct payments for farmers; oilseeds and milk are two other sectors affected by the new reform.

Conclusion

The mechanisms of the CAP are based on internal market support instruments, coupled with export refunds enabling EU farmers to sell their produce on the international market at a profit in spite of the lower prices prevailing abroad. Although the policy has been successful in enabling regular supplies of food products in the EU, it has been criticized for swallowing too large a portion of the community finances. Trade frictions with the EU's main agricultural partners, environmental concerns, high costs to developing nations as well as an unfair distribution of EAGGF benefits amongst EU farmers have also been decried. The need to alter significantly the generous (and onerous) price system embodied in the CAP, and to favor structural change in the European agricultural holdings was clearly perceived during the 1960s. The Mansholt Plan of 1968 addressed these issues for the first time. In spite of increasing budgetary problems, very little progress was actually made until 1984. Sporadic initiatives took place between 1984 and 1991, with the implementation of the first alternative rural development strategies. Drastic changes occurred with the 1992 CAP Reform, which fundamentally modified the price support mechanism of the CAP through tariffication and a substantial decrease in support prices. Indubitably, the multilateral trade negotiations conducted under the auspices of GATT had an accelerator effect on the process of CAP reform. The CAP is today a more market-oriented policy, despite its re-distributive aspect represented by large compensation payments. More importantly, the 'accompanying measures' with their environmental and social dimensions are now an integral part of the new market organizations in the EU. The 1992 reform of the CAP has been successful in halting the escalation of EAGGF guarantee expenditure since the mid-1990s. In addition, intervention stocks have decreased, and this trend has only been temporarily jeopardized in the beef market by the outbreak of BSE in the mid-1990s. Cereal stocks have shrunk, with the exception of rye, and domestic (producer) prices are now close to world market prices, with the exception of the beef sector where effective support price levels are still high when compared with world market prices (CEC, 2002b). It should be noted that in the cereals sector, support levels are 'decoupled' from production, whereas in the beef sector they are still paid per head of animal (which represents an incentive to produce up to the payment ceiling). Although, at the international level, the CAP is now more compatible with the principles of free trade than it was ever before, the policy is still fraught with high costs, which are borne mostly by the average European consumer. Clearly, the reordering of the EU agricultural markets in the 1990s has meant that the nature of support has changed. EAGGF expenditure has evolved from supporting the high domestic market prices to supporting directly the farming community, through direct compensation schemes, with little positive effects for the European consumer. What needs to be assessed is whether the savings associated with the depletion of stocks in the EU have reasonably offset the higher costs induced by the compensatory payments. The money saved, as a result of freer trade, seems marginal as the CAP still swallows nearly half of the EU budget.

The principle of strict management of markets is central to further CAP reforms, as necessitated by the enlargement to East-European countries, characterized by large and inefficient agricultural sectors. Agreed by the 1999 European Council in Berlin, the Agenda 2000 reform of the CAP is an important step in the reform process, and it reiterates the necessity to reorder the EU agricultural markets, so as to allow it to become more in line with market forces.

The CEC (2002) document acknowledges the fact that 'further steps are necessary to improve the market orientation of European farming' (p. 6), in particular in terms of price differentiation.

Key terms and concepts

SPAAK report
Target price
Intervention price
Export refunds

Green money
EAGGF
Self-sufficiency ratios
Mansholt plan
1992 CAP reform
Market-oriented policies
Beef crisis and BSE
Agenda 2000

Notes

1 A quarter of Italian exports were coming from farms at that time.
2 The value of the unit of account was 0.88867088 grams of gold. This weight was roughly equivalent to that of the US$ during the Bretton Woods era (1944–71).
3 The reference period was initially 1969–73.
4 The EMS pledged to ensure 'a zone of monetary stability in Europe' by allowing the participating currencies to fluctuate within an extra limited margin of fluctuation (±2.25% from the bilateral rate; ±15% since August 1993). Note that the ECU served as the basis for the definition of the Common European Currency, the EURO.
5 In the European literature, this fund is often referred to as the FEOGA, which is the French translation of EAGGF (that is Fonds Européen d' Orientation et de Garantie Agricole).
6 Before 1971, the financing of the EEC was drawn from member states financial contributions. The most thorough description and analysis of the EC/EU Budget are contained in Strasser (1992).
7 Note that the agricultural levies were introduced when the CAP was born. They were taken into account for the calculation of the member states contributions prior the 1971 budget.
8 The Uruguay Round started at Punta del Este in 1986, and it was scheduled to last only four years. This has been the last round of multilateral trade negotiations conducted under the auspices of the GATT (General agreement on Tariffs and Trade). The GATT has accomplished a considerable worldwide tariff decrease since the late 1940s, and many non-tariff impediments have also been eliminated as well.
9 The term 'subsidy equivalent' appeared first in the Treaty of Rome.
10 For a detailed presentation and mode of calculation of PSEs see Chapter 8.
11 The marketing year for all agricultural products runs from the 1 July to 30 June of each year.
12 Another implication of the 'correcting factor' is that positive MCAs are converted into negative MCAs. This conversion is known as the 'switch-over' mechanism.
13 That is, a 10 per cent official devaluation of the Irish Punt, combined with the devaluations of the Lira (6.68%), of Sterling (1.19%), and of the Drachma (2.79%), in November 1992. These devaluations caused the other currencies to appreciate slightly.

Bibliography

Balassa, B. (1988) 'Agricultural policies and international resource allocation', *European Review of Agricultural Economics*, 15: 2/3, 159–71.

Baudin, P. (1991) 'La fixation des prix agricoles pour 1991/1992', *Revue du Marché Commun et de l'Union Européenne*, No. 349, Juillet/Août, 535–51.

Cahill, C. and Legg, W. (1989–90) 'Estimation of agricultural assistance using producer and consumer subsidy equivalents: theory and practice', OECD Economic Studies: Special Issue, *Modelling the Effect of Agricultural Policies*, No. 13, Winter, 13–43, Paris: OECD.

CEC (1988) *The Future of Rural Society* (Bulletin of the European Communities, Supplement No. 4, Commission Communication to Parliament and the Council).

CEC (1991a) *The Development and Future of the CAP* – Reflections Paper of the Commission (COM (91) 100 final, 1 February, Brussels and Luxembourg).

CEC (1991b) *The Development and Future of the Common Agricultural Policy* – Proposals of the Commission (COM (91) 258 final, 22 July, Brussels and Luxembourg).

CEC (1993) *The New Regulation of the Agricultural Markets* – Vademecum (Green Europe: Luxembourg).

CEC (1995) *The Agricultural Situation in the European Union* (1994 Report, Brussels and Luxembourg).

CEC (2002a) *The Agricultural Situation in the European Union* (2000 Report, Brussels and Luxembourg).

CEC (2002b) *Mid-Term Review of the Common Agricultural Policy* (Communication from the Commission to the Council and the European Parliament, COM (2002) 394 final, Brussels and Luxembourg).

Conway, A.G. and O'Hara, P. (1985) *Integrated Rural Development: an Irish Case Study to Identify Strategies for the West Region* (Economic and Rural Welfare Centre, An Forás Taluntais, Dublin).

Duby, G. and et Wallon, A. (1976) *Histoire de la France Rurale* (Tome 3, de 1787 à 1914) (Paris : Editions du Seuil).

Duchêne, F., Szczepanik, E. and Legg, W. (1985) *New Limits on European Agriculture: Politics and the Common Agricultural Policy* (London: Croom Helm).

European Commission (1994a) *EC Agricultural Policy for the 21st Century* (European Economy, No. 4, Brussels and Luxembourg).

European Commission (1994b) *The Economics of the Common Agricultural Policy (CAP)* (European Economy, No. 5, Brussels and Luxembourg).

European Parliament (1994) *Fact Sheets on the EP and the Activities of the European Union* (DG for Research, Luxembourg).

European Report (1996) *Beef: EU Farm Ministers in Urgent Search for Crisis Support Cash* (No. 2160, 25 September, IV-7–8, Brussels).

Koester, U. and Tangermann, S. (1976) *Alternative der Agrarpolitik* (Münster: Hiltrup).

Le Monde (1995) 'Zéro stock et frigos vides, la Communauté verse en Ecus son aide aux démunis' (article by E. Fottorino, 26 September, Paris).

Marinelli, A., Casini, L. and Romano, D. (1990) 'Valutazione economica della ricreazione', *Genio Rurale*, No. 9.

Matthews, A. (1995) 'Agricultural competitiveness and rural development', in O'Hagan, John W. (ed.) *The Economy of Ireland – Policy and Performance of a Small European Country*, 328–62 (Dublin: Irish Management Institute).

OECD (1992) *Agricultural Policies, Markets and Trade, Monitoring and Outlook* (Paris: OECD).

OJEC (1992) *Council Regulations of June 1992* (L180, Vol. 35, Brussels).

O'Malley, E. (1992) *The Pilot Programme for Integrated Rural Development 1988–1990* (Economic and Social Research Institute, Dublin).

Pareto, V. (1889) Atti del Congresso di Roma per la pace e per l'arbitrato internazionale, Maggio 1889, Città di Castello.

van Riemsdijk, J.F. (1972) 'A system of direct compensation payments to farmers as a means of reconciling short run to long run interests', *European Review of Agricultural Economics*, 1: 2, 161–89.

Roningen, V.O. and Dixit, P. (1989) *Economic Implications of Agricultural Policy Reform in Industrial Market Economies* (US Department of Agriculture, Economic Research Service, Washington DC).

Strasser, D. (1992) *The Finances of Europe* (Commission of the European Communities; European Perspectives, Luxembourg; 3rd English edition).

Tarditi, S. (1984) 'La crise de la PAC: un point de vue italien', *Economie Rurale*, Vol. 163, 27–64.

Tracy, M. (1989) *Government and Agriculture in Western Europe 1880–1988* (3rd edition, New York, London: Harvester Wheatsheaf).

World Bank (1986) *Agricultural Policies in Developing Countries – Exchange rates, prices and taxation*. World Development Report, Chapter 4, 61–84 (New York: Oxford University Press).

Yeats, A.S. (1981) 'The influence of trade and commercial barriers on the industrial processing of natural resources'. *World Development*, No. 9, 485–94.

5
Technology and Innovation

Objectives of this chapter

- To discuss the relationship between technology and economic growth.
- To review briefly the major innovations in agriculture.
- To introduce the dominant models of innovation diffusion and adoption.
- To highlight the role of innovation in agricultural development.
- To analyze agricultural innovation in the EU, with an emphasis on the EU technology programs, and extension services on EU countries.

Introduction – technology and economic growth

This chapter is devoted to the analysis of technological change in agriculture. The central role of technological change and of innovation in agricultural production can be demonstrated with basic statistical facts. Between the years 1950 and 1998, the world population more than doubled, and yet global per capita grain production increased by 15 per cent; during the same time, world per capita harvested acreage declined by 50 per cent (Zilberman, 1997). Because of the crucial role played by technological change in Europe in the last 50 years, this chapter starts with a discussion on the relationship existing between technology and growth.

The causes of economic growth have been the theme of numerous debates and writings by many economists, historians, sociologists and philosophers alike since at least the time of mercantilism. Those, like Ellis (1826), interested in the application of scientific discoveries to the production of goods in both the agricultural and manufacturing sectors, could not fail to make the association between increased knowledge (or new technology) and macroeconomic growth in 19th century Great Britain. Surprisingly, despite the triumph of science and technology for more than two centuries, the production function used by economists remained remarkably static, with its two factors of production, capital and labor, and deprived from any qualitative attributes. In the early part of the 20th century, the link between technology and economic growth starts being strongly asserted in the writings of both Kuznets (1930) and Schumpeter (1912, 1943). In particular, Kuznets described how the output of any given good tends to follow an S-shaped pattern over time, and how this pattern is affected by the rate of technical progress in this particular industry. Stressing the importance of inventions and knowledge creation in economic phenomena, Kuznets concluded that the growth rate of an industry declines with its inventive potential.

From modern growth to 'new' growth theories

The 1950s were a very fruitful decade in the field of economic growth theory. The publication of groundbreaking conceptual as well as empirical studies by Fabricant (1954), Abramovitz (1956) and Solow (1957) contributed to shed some light on the famous 'residual', that is that portion of output growth that could not be explained by the two standard factors of production of the neoclassical production function. In particular, Abramovitz suggested a few pertinent avenues that proved helpful in trying to explain this 'residual', which in his eyes was simply a 'measure of our ignorance about the causes of economic growth' (Abramovitz, 1956: 11). He suggested for example to broaden the notion of capital to include expenditure on health, education and research, and he advocated the construction of a more adequate index of labor input, in which the more highly educated workers would be accorded a greater weight. The following year, Solow's study provided the first quantitative analysis of the contribution of technology upgrading to increases in American output. Assuming constant returns to scale, Solow uses a new aggregate production function of the form $Q = f(K, L; t)$, where the variable t refers to 'any kind of shift' in the production function, including educational improvements of the labor force. Applying this model to the case of US non-farm output between the years 1909 and 1949, he concludes that more than 87 per cent of the increase in gross output per man hour during the period is attributable to technical change (Solow, 1957).

These various studies led to a renewed interest in dynamic models of growth in the 1960s, with publications by Arrow (1962), Uzawa (1965), Phelps (1966) and others. Although the assumption of increasing returns to scale features in Arrow's work, it is absent in Uzawa's optimizing growth model in which nevertheless human capital is seen as a core variable. A clear distinction between endogenous and exogenous technical change appears in the work of the Dutch economist Heertje (1977). Providing an extremely thorough – and very much neglected – review of contributions leading to the emergence of what was subsequently referred to as the 'new' growth theory, Heertje defines endogenous technical change with due reference to: '(i) long-term changes in the ratio between the prices of the factors of production; (ii) learning processes concerning production; and (iii) investment in education and research' (Heertje, 1977: 174). This is in contrast with standard growth theory that views technical change as being simply embodied in capital goods. In standard production theory, technical change is indeed simply identified as a shift in the production function, without any consideration for the factors that influence or explain this shift.[1] In summary, the modern growth theories of the 1960s and 1970s see human capital accumulation, learning considerations, information and knowledge as essential factors in the process of economic growth. In particular, education and knowledge produce positive externalities or increasing returns.[2]

It is on the solid foundations of this rich economic literature that the so-called 'new' growth theories emerged in the 1980s. These theories have been commonly associated with the names of Romer (1986) and Lucas (1988) who provided some refinements on earlier models. The field of agricultural economics did not escape from the authoritative influence of endogenous growth models. Evenson (1990) provides a synthesis of early studies using modern growth concepts in agriculture.

In a world where the interdependency between nations is as pronounced as nowadays, a new technology, idea or knowledge emerging in one country or region of the world, can diffuse successfully to other regions, through various means, in particular through trade and/or foreign direct investment. The importance of knowledge diffusion, as espoused in early models, is explored in Section 5.2, whereas newer models stressing the importance of extension services are presented in Section 5.3. Diffusion to the benefit of laggard economies or of the less well off farmers, when analyzed at the microeconomic level, is the main idea behind structural funding and technological collaboration in the EU (Section 5.4). This section will also conclude with a brief discussion on the return to agricultural research investment. Before touching on the issue of agricultural innovation in the EU *per se,* we propose a brief review of the major innovations in European agriculture (Section 5.1).

5.1 Major innovations in European agriculture

The way innovations and technical change have affected agriculture since classical antiquity may first be appraised by a synopsis of economic growth trends before and after the industrial revolution that started in the late 18th century. What Table 5.1 infers is that, since the era of industrialization and mechanization (1820) in the world, the pace of knowledge and of technological change increased rapidly, bringing about substantial macroeconomic growth and unprecedented innovative opportunities for the agricultural sector. The important growth in the population during the period reflects indeed substantial nutritional improvements made possible by increased agricultural output, which in turn is attributed to better farming methods.

During the first thousand years that followed the collapse of the Roman Empire, progress in terms of population growth was minuscule; with regard to per capita product, improvements were almost negligible. However, this is not to assert that inventive output was inexistent prior to the industrial revolution. On the contrary, even during classical antiquity and medieval times, major inventions, technical improvements and innovations were brought to the fore, which did revolutionize the agricultural productive methods. Clearly, the innovations in the fields of mechanical engineering, physics and chemistry during and after the industrial revolution led to exponentially increasing improvements in the area of food production.

Table 5.1 Average annual compound growth rates during three eras

	Population	Per capita GDP
0–1000	0.00	−0.01
1000–1820	0.20	0.14
1820–1998	0.60	1.51

Source: Derived from Maddison (2001).

In this section we provide an overview of the major innovations affecting the agricultural sector since classical antiquity. What we propose is merely a very rough overview spanning from the invention of the wheel and irrigation techniques to contemporaneous breakthroughs in the biotechnology field.[3]

5.1.1 Classical antiquity

Compared with what was to follow, the period spanning from 500 BC to 500 AD brought only few innovations which were directly applicable to agriculture. One such innovation was the lever attributed to Archimedes, which, when combined with the helix, could be used for pressing. Representing an example of this new technology, the winepress is believed to have appeared at around 70 AD (Mokyr, 1990). The waterwheel was another significant invention made during the first century AD. Waterwheels were used for flour milling. Improving on the hydraulic engineering techniques of the Egyptians, the Romans were able to apply them for irrigation and drainage purposes. As the Celts and Gauls were more concerned about labor saving techniques than the Romans, many inventions and innovations in the area of agricultural equipment and techniques have been credited to these two people.

5.1.2 Medieval times

During the 'dark' age, spanning from the collapse of the Roman Empire (that is, end of the fourth century) to the beginning of the 12th century, European agriculture benefited from a number of substantial innovations. These were the heavy plow in the sixth century, the horse collar, and the idea of the three-field system of crop rotation. This new farming system enabled the cultivation of additional crops. Other important innovations during that first part of medieval times were the wind mill, introduced in the 12th century, the nailed horseshoe and the modern horse collar which permitted the increased use of horses in the agricultural sector, with ensuing productivity effects. The 'agricultural revolution' experienced by the Muslim world between the 8th and the 12th centuries enabled the introduction of many Asian products which started to diffuse to Europe. This is how for example sorghum and rice, hard durum wheat, oranges, bananas, watermelons, asparagus and spinach, became gradually part of a substantially improved European diet (Watson, 1983).

During the second part of the middle age (between 1200 and 1400), many of the European inventions emerged in other areas, such as in metallurgical engineering, ship construction and mechanical engineering, preparing Europe for technological leadership and sowing the seeds of modern global capitalism. For example, ship construction and navigational techniques improved dramatically, preparing European explorers to venture into the unknown southern hemisphere. The inflow of new capital into the agricultural sector generated important changes in the sector. The introduction of the silk worm in Europe during this time allowed the emergence of a new burgeoning silk industry, in both what is known today as France and Italy.

5.1.3 Renaissance

During the Renaissance period (starting at around 1400), the rate of population growth increased markedly, compared with the previous era, with small increases in per capita GDP growth. This period is characterized by ingenious ideas and discoveries owed to renowned scientists such as Leonardo da Vinci and Copernicus. Mokyr (1990) sees this period as being as inventive as the period of the industrial revolution, although the impact of many inventions failed to translate successfully into viable innovations. In Holland, the 'new husbandry' system in agriculture, that is the elimination of fallowing and the introduction of new crops and of stall-feeding of cattle, led to important increases in productivity, and to large improvements in human well-being, at least in the long term. However, it is believed that the new techniques introduced in the agricultural sector at the time were more capital and land saving than labor saving (Mokyr, 1990). Technical progress at the time greatly increased the supply of fisheries products, an essential source of protein. In particular, the improved techniques of fish catching introduced by the Dutch fishermen around 1415, helped the Dutch fishing ships establish domination in the North Sea.

5.1.4 The Industrial revolution and after

The period spanning from 1750 to 1820 witnessed the take-off of the industrial revolution, which is conventionally thought to have diffused from Britain to other countries of Europe, including czarist Russia. During this time, and in spite of often appalling employment and living conditions of many laborers, both the population and per capita output growth rates doubled, when compared with the previous period. These trends were to defeat the pessimistic predictions of Robert Malthus who wrote his *Essay* during that time. After 1820, per capita GDP increases eightfold, reflecting the jump in technological improvement in continental Europe and in Britain, and leading to substantial productivity growth.

The industrial revolution was paralleled with important structural changes, as can be seen from Table 5.2. The share of the workforce employed in the agricultural sector falls sharply from the middle of the 19th century in Europe, except for the southern European countries where protectionism was more prevalent. As was evoked in Chapter 4, protectionism tends to halt positive adjustments in the agricultural sector. In the latter part of the 19th century, the need to devise immediate solutions in a context of crisis led some European Governments to overemphasize price support policies at the expense of structural measures. As a result, technological progress slowed down in these countries. In contrast, as they held to the *laissez-faire* doctrine before the First World War, the United Kingdom, Denmark and The Netherlands were able to promote substantial structural adjustments in their agricultural sectors. In Britain, for example, non-viable farms disappeared and the large units were able to achieve high productivity rates. During the same period of time, the structure of the French protectionist agricultural sector was virtually unchanged.

Innovations in the chemicals, machine tool and transport industries had major impacts on the production, preservation, storage and transport of food. Table 5.3 shows how innovations in the chemicals industry have diffused to the agricultural sector. Again, the north–south divide appears clearly with extremely high levels of fertilizer utilization in a country such as Holland, where land is scarce.

5.1.5 Modern times – new farm techniques and the 'biorevolution'

The European agricultural landscape observable at the beginning of this new millennium is in perfect contrast with that of immediate post-war

Table 5.2 Share of the workforce engaged in agriculture, 1800 to 1920/30

	1800	1850	1900/10	1920/30
United Kingdom	36	22	9	6
The Netherlands	na	44	28	22
Germany	na	55	28	23
Belgium	na	48	32	23
France	55	52	42	35
Italy	na	na	61	55
Spain	65	64	65	51

na: not applicable.
Source: Simpson (1995: 18).

Table 5.3 Utilization of chemical fertilizers in Europe, 1913 and 1932

	Phosphates		Potash		Nitrogen	
	1913	**1932**	**1913**	**1932**	**1913**	**1932**
Holland	1.99	1.98	6.74	5.61	1.18	1.71
Belgium	0.90	0.90	1.60	3.66	1.83	1.31
Germany	0.58	0.65	4.87	7.99	0.65	1.12
Great Britain	0.97	0.82	0.43	2.01	0.64	0.63
France	0.46	0.49	0.22	2.16	0.23	0.51
Italy	0.35	0.31	0.06	0.16	0.11	0.29
Spain	0.11	0.20	0.09	0.08	0.04	0.28
Greece	–	0.04	–	0.02	–	0.02

NB: A figure of 1.00 implies that 'adequate' quantities were applied per hectare.
Source: Simpson (1995: 110).

Europe. Drastic changes have occurred after the Second World War, particularly in southern Europe, where the intensification of production methods has been widespread. Major modern farm techniques, such as irrigation technologies have permitted the increased application of chemical fertilizers across EU agricultural holdings. Mechanization, allowing the substitution of capital for labor, and improvements in the quality of human capital, associated with formal schooling have led to an increase in yield per hectare, as shown in Table 5.4. All these new production methods have facilitated product innovation and product differentiation.

During the last quarter of the 20th century, environmental innovations permitting the use of techniques and methods to improve land management rather than to increase farm productivity were promoted. Aimed at 'sustainable agricultural development',[4] these modern environmental innovations allow to put an end to soil degradation with the loss of the productive potential of farmland for future generations. The potential loss inflicted to future generations is not only one in terms of productivity, but it concerns also the damage to the wildlife habitat, to ecological diversity, and the exhaustion of non-renewable farm inputs. In line with the polluter pay principle, the costs of environmental innovations may be borne by the individual farmer, while the benefits are social.

It is worth remembering that the developing world, and in particular Asia, is the region *par excellence* where post-war agricultural innovation has had the most visible and remarkable effect. Since the late 1960s, the green revolution consisted in introducing chemical fertilizers, increased

Table 5.4 Average wheat yields in Europe, 1909/13 and 1961/65

	Tons per hectare	
	1909/13	**1961/65**
Denmark	3.3	4.1
Holland	2.4	4.4
Belgium	2.5	3.9
UK	2.1	4.0
Germany	2.4	3.2
Switzerland	2.1	3.3
Ireland	2.5	3.3
Sweden	2.1	3.4
Norway	1.7	2.6
Austria	1.4	2.6
France	1.3	2.9
Italy	1.1	2.0
Finland	1.1	1.7
Greece	1.0	1.5
Spain	0.9	1.1
Portugal	0.7	0.8

Source: Simpson (1995).

capital investment and irrigation methods, as well as high-yielding varieties of cereals, which proved superior in terms of yield potential and resistance to pests. This involves the technique of plant breeding, which consists in crossing selectively two parent plants to produce offsprings of superior quality (for example resistant to diseases). The spectacular increases in agricultural output region-wide did help many Asian nations to embark upon an almost uninterrupted cycle of economic growth during the 1970s and 1980s. Acclaimed for its success, the green revolution has nevertheless been severely criticized for nurturing inequality, social antagonism and unrest in these countries (Santikarn Kaosa-ard and Rerkasem, 2000).

One major driving force of agricultural innovation in Europe and elsewhere nowadays is the 'biorevolution' that countries are slowly but inexorably experiencing, including several countries in the developing world (see Box 1). Borrowing the definition of the ADB (2001: 10), biotechnology can be defined as 'any technique that uses living organisms, or parts of such organisms, to make or modify products, to improve plants or animals, or to develop microorganisms for specific use'.

Box 1 The emergence of the 'biorevolution'

In the early 1970s scientists started developing new techniques for precise recombination of portions of deoxyribonucleic acid (DNA) and for transferring portions of DNA from one organism to another. These techniques are referred to as genetic engineering. The expressions 'transgenic material', 'living modified organisms', 'genetically engineered organisms', 'genetically modified organisms', and 'genetically improved mechanisms' are used indiscriminately as they are all synonymous. The first transgenic species appeared in the 1980s, with the first transgenic plant in 1982 and the first transgenic pig in 1986. The genetically modified organisms (GMOs), which have received so much attention, particularly in Europe, are one among other key components of biotechnology, alongside genomics, and vaccine technology which is aimed at controlling livestock and fish diseases. In particular, research in the area of genomics, defined as the molecular characterization of all genes in a species, should greatly help in identifying for example the plants' genes controlling the resistance to certain insects, the tolerance for soil toxicity, and so on. Plant genetic engineering helps at improving the nutritional quality of crops, by increasing for example their specific vitamin contents, and by reducing fat components, and so on.

Through genetic manipulations, biotechnology techniques are thought to provide novel ways to improve plants and animals; moreover, they enable to produce new products with which to treat plants and animals. Biotechnology applications require intensive research efforts, and most of the research conducted in the field is in the hands of a few large private-owned multinationals, given the predominant *geistzeit* since the 1980s, which has been one of minimal governmental intervention in economic affairs. As will be seen below, the EU sponsored programs in the biotechnology field, introduce nevertheless an exception to this rule (see Section 5.4). The distinction between private and public research is worth mentioning particularly in the background of what has happened in the past. The green revolution technologies were developed essentially for the public, and by public research institutions and philanthropic foundations. By contrast, innovations in the field of biotechnology are mostly the prerogative of competing private businesses.

As for any new scientific knowledge introduced so far, but perhaps even more now than ever before, biotechnology methods generate both positive as well as negative consequences. Although the 'biorevolution' is aimed at bringing new solutions to the problems of food security, of natural limits on food production imposed by pollution (such as soil erosion), and eventually of poverty, it also brings many unanswered questions in terms of environmental and food safety concerns. Although the theoretical benefits of GMOs are known, it seems that the first generation of GMOs has not yet benefited the final consumer, let alone the final consumer in poor countries. According to the ADB (2001), the benefits so far would have accrued to the producers of inputs, such as chemical companies, seed producers, and perhaps to farmers. Among the many concerns voiced about the development of GMOs are the following main issues:

(i) *Food safety*: The risks in terms of human health are a controversial issue; some studies have provided evidence of toxic reactions and of fatal food allergies generated by genetically engineered products.

(ii) *Environmental concerns*: Although there does not seem to be, among scientists, agreement on whether the production of GMOs will increase or rather decrease pesticides residues in soils, a major issue here is the genetic pollution generated by GMOs. The DNA of non-genetically engineered and organic crops could be contaminated through birds, insects, rain and wind carrying genetically modified pollen from fields planted with GMOs. Also, the genetic evolution of species and the easiness with which they adapt to new circumstances, could lead to the creation of superweeds, superpests, new bacteria and viruses.

(iii) *Socio-economic issues*: Biotechnology research and production is concentrated among a few large business entities in the developed world. In 1997/98, the top ten firms in the world[5] accounted for between 80 and 90 per cent of the world market of agro-chemicals (CEC, 2001a). These firms have so far concentrated their research efforts on the most economically viable products, such as the temperate crops for large farmers in developed countries. This may increase the gap between rich and poor countries, defeating one of the main purposes of biotechnology research, which is, supposedly, to provide adequate food supplies to a growing population, mostly in poor countries. Moreover, farmers depend increasingly on a smaller number of suppliers for crop protection, increasing thereby their dependence and their vulnerability.

Finally, research conducted on the first generation of GMO crops in the USA has failed to bring evidence of any increases in both profitability and yields for farmers. On the contrary, studies on herbicide tolerant soybeans and insect resistant corn have concluded that GM crops show lower yields than non-GM crops in most field trials under investigation. Also, the introduction of GM seeds has so far resulted in higher input costs for farmers (CEC, 2001a). This latter concern may however be reversed in the future, for the price of new products tends to decrease over time, once research and development costs have been recovered and once economies of scale have set in.

(iv) *Cultural and ethical issues*: The introduction of GMOs in some of the most renowned European cuisines is, in the eyes of many European consumers, simply unthinkable; finally, on an ethical level, it suffices to add that the fabrication of the first genetically modified monkey in 2000 creates the risk of its replication in human beings.

In short, there is a lot of scientific uncertainty surrounding the development of genetically engineered food products.[6]

Finally, to conclude the section, we should note that all these different technical changes and new technologies are connected in some way. As pointed out by Anderson and Herdt (1988), the emergence and rapid adoption of many new crop varieties in the world were significantly influenced by the availability of irrigation and of mechanization. More significantly, the intensified use of some fertilizers (such as nitrogenous fertilizers) became much more profitable with the new varieties. There is indeed a complementarity between the development of high-yielding new varieties and fertilizer inputs.

5.2 Models of innovation diffusion and adoption

Following the conceptual developments proffered by the moral philosophers of the 19th century, such as Charles Babbage (1832), the Austrian-born economist Joseph Schumpeter made clear the distinction between invention, innovation and diffusion (Schumpeter, 1912, 1943). Invention is purely an intellectual act for it refers to the creation of a new idea, whereas innovation entails the transformation of the idea into practical use by a business entity.

The distinction above is essential, given that all inventions do not systematically translate into innovations, and that all innovators are not necessarily inventors. Indeed most inventions fail to be converted into exploitable business activities, as was the case for many brilliant ideas born during the period of the Renaissance. Depending on the industry, the time lag between the invention of a new product (or process) and its introduction onto a market can span from a few weeks to several years, even decades.[7]

The evidence accumulated in most post-war economies, that is during a time of greater international economic integration and of scientific discoveries facilitating the dissemination of knowledge, such as the advances made in the area of telecommunications, suggests that innovation is normally associated with diffusion. This dissemination can take place within a country's industrial sector proper, that is the agricultural sector, allowing all firms to become eventually *au fait* with the new knowledge. It can also take place on a spatial level by enabling the diffusion of innovations across regions and countries. Interestingly, the agricultural sector was one of the few areas, together with the medical and educational spheres, that stirred up interest in the diffusion of innovations during the immediate post-war period. Diffusion is also the prerequisite for imitation, which is simply the copying and improvement of the innovation by others.

The way in which diffusion takes place is a complex phenomenon that transcends the boundaries of economics. There are indeed many cultural, historical as well as societal aspects that render diffusion a successful process, particularly on the part of the recipient firm or country; concomitantly, the non-economic factors mentioned above may hamper a successful diffusion either spatially or within an industry. Indeed, for an innovation to diffuse and to be absorbed successfully into an economy or by a firm, a number of conditions must be met. At the macroeconomic level, even if all the economies have access to the same technology, national growth rates can still differ, depending on the quality of human capital and on the incentives to adopt the new technology in the recipient economies. The ability for a recipient economy to absorb an innovation and to let it diffuse, is referred to as 'absorptive capacity' by Cohen and Levinthal (1990). Technology transfer and spillovers are the two vehicles by which diffusion occurs. There is a luxuriant literature on technology transfer and on spillovers, and the objective of this chapter is not to reconsider the main highlights of this literature. Suffice to state that the main difference between the two concepts is that technology or knowledge transfer is a well-planned and intentional process, whereas spillovers can be defined as hypothetical knowledge gains that may (or may not) diffuse to the recipient firm or country.

As a consequence, the perfect knowledge assumption embraced in Léon Walras' neoclassical theory (Walras, 1874), implying that diffusion and imitation will be instantaneous, appears obsolete. However, before we turn to a more modern analysis of innovation and diffusion models in the next section, we first venture into the neoclassical based models that have dominated agricultural economic thinking for many decades.

There are two major such models explaining technological change and diffusion in agriculture: Griliches model of innovation diffusion and adoption (Griliches, 1957), and Hayami and Ruttan's model of induced innovation (Hayami and Ruttan, 1970).

5.2.1 The model of innovation diffusion and adoption

Griliches' pioneering article aimed at throwing light on the way in which an innovation propagates across different regions of the USA. Although in the first version of this model, Griliches acknowledges the fact that knowledge diffusion is not instantaneous, this model is very much rooted onto neoclassical thinking given for example that farmers behave in line with the profit maximization objective. Several refinements of Griliches' model in the 1960s and beyond have made it the most dominant model of innovation diffusion. Among the features of this model is the assumption that the individual decision-making process allows the adoption of an innovation to be seen as a complex set of stages including: awareness, information, evaluation, trial and adoption. This, in turn, leads to the classification of farms into several categories such as early adopters, early majority, majority, late majority and laggards (Rogers, 1983).

A standard model of innovation diffusion, known also as the 'epidemics model', can be presented as follows. Suppose that $x(t)$ is the fraction of potential adopters (or farms) who adopt a given innovation at time t. The rate of diffusion will be given by:

$$\frac{dx(t)}{dt} \tag{5.1}$$

This rate of diffusion is proportional to the fraction of adopters $x(t)$ in time period t, as well as to the fraction of potential adopters left that is: $[1 - x(t)]$. As time elapses, $x(t)$ rises and $[1 - x(t)]$ falls, suggesting that there is a greater likelihood that a non-adopter will come into contact with an adopter, and will himself adopt the innovation. Consequently, we can write:

$$\frac{dx(t)}{dt} = \beta \,.\, x(t) \,.\, [1 - x(t)] \tag{5.2}$$

where β, constant, refers to the coefficient of adoption/diffusion. This coefficient measures the speed at which the innovation diffuses across farms. Equation 5.2 has the following solution:

$$x(t) = \frac{1}{1 + \exp(-\alpha - \beta t)} \tag{5.3}$$

Equation 5.3 is the expression of a logistic curve, showing that the rate of diffusion increases until an inflexion point, and then slows down. The logistic curves represented in Figure 5.1 depict the rate of diffusion/adoption of hybrid corn by farmers across American regions, as presented in Griliches' work (Griliches, 1957).

This model is useful in that it enables to make various estimations on the adoption/diffusion coefficient β, and to relate these to various micro, meso or macroeconomic variables. For example, it has been found that this coefficient increases with the increased profitability of the innovation (Gaffard, 1990). Other critical factors in influencing the coefficient of diffusion are firm size and quality of the management. Although Griliches accepts the limited applicability of his model, he nevertheless highlights the fact that 'it must be possible' to apply it to the case of the use of fertilizers and machinery in US farms (Griliches, 1957: 521). Several studies based on Griliches' seminal work have shown the spatial diffusion and adoption of new discoveries in the area of agricultural economics; this has been the case for the following

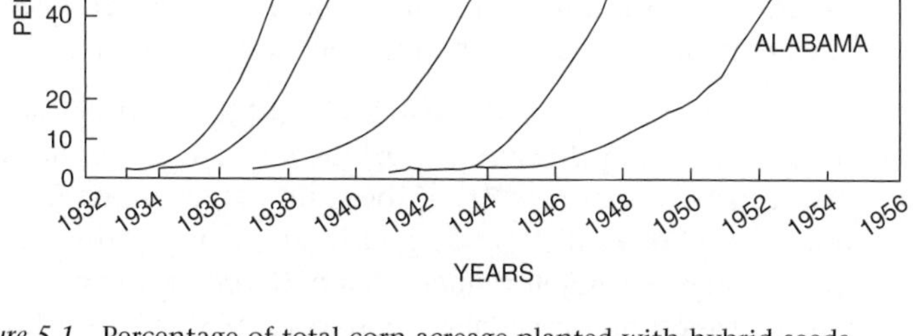

Figure 5.1 Percentage of total corn acreage planted with hybrid seeds
Source: Griliches, 1957 (USDA, Agricultural Statistics, various years).

commodities: hybrid sorghum, poultry, sugarcane, wheat, maize, cotton, tomato, rice and soybeans (Evenson, 1990).

Although, relatively successful in the 1960s, this model of innovation diffusion has been the subject of many criticisms in subsequent decades. There are a number of limitations that have been pointed out in the literature (Coombs *et al.*, 1987; Gaffard, 1990; Vanclay and Lawrence, 1994). The major limitations and criticisms can be summarized as follows:

(1) The model is static in that it describes the diffusion of a given innovation over time in an environment that does not change. Evidence shows on the contrary that the innovation/diffusion process is an interactive phenomenon as many innovations are subject to incremental changes during the course of their diffusion, making the population of potential adopters increase over time. Therefore, a more appropriate model of diffusion would be represented by an envelope of logistic curves, each reflecting the specific characteristics of the evolving innovation and environment in which the innovation diffuses. As Gaffard (1990) rightly points out, this envelope may not assume the form of a logistic curve.

(2) The model gives too much weight to the demand side, that is to the adopters of the innovation, neglecting totally the profitability prospects of those supplying the innovation. Again, it is clear that for the innovation to diffuse, it must be profitable to both adopters and producers. Also, the model has become obsolete in that it is not suitable to explain the diffusion of modern conservation technologies or of environmental innovations. New environmental technologies are more complex than other commercial innovations, and they may therefore provoke a greater resistance to adoption. It has been shown that farmers are more likely to adopt environmentally compatible techniques when they feel personally at risk from environmental degradation (Rickson *et al.*, 1987). Finally, environmental innovations tend to be indivisible, and cannot therefore be adopted in parts, negating the possibility of trial.

(3) Finally, Vanclay and Lawrence (1994) highlight the many social and crucial issues neglected in this model, such as the variables underlying the unequal distribution of benefits of the technology among farmers. In his pioneering work, Griliches was greatly suspicious of the importance of the sociological variables, since most of them, in his eyes, could be 'redefined [...] as economic variables' (Griliches, 1957: 522). Although the importance of non-economic variables (that is sociological, historical

and cultural) may have not been very pronounced in the case of innovation diffusion across states in the USA of the 1950s, these non-economic variables have assumed such an importance in modern times, that they do matter, particularly in the case of a set of culturally heterogeneous countries such as the one represented by the EU. For example, the inclusion of sociological factors, such as the different approaches to education amongst farming communities in Europe, may explain why the services provided by the 'extension' (or advisory) agencies seem to reach certain groups of farmers first.

5.2.2 The induced innovation model

The model of induced innovation considers the importance of changes in relative factor prices as inducing technical, and even institutional change. According to this model, the nature of technical change in the agricultural sector represents a response to long-run trends in relative factor prices, which are in turn explained by factor endowments and product demand growth. The model views changes in factor prices as essential, for an increase in the price of one factor, relative to others, will induce technical change that reduces the use of that factor relative to the other factors of production. Developed by Hayami and Ruttan (1970, 1985), the model explains therefore the mechanisms by which a society choses a given trajectory of technical and institutional change in agriculture. It should be noted that the idea according to which a change in relative factor prices can explain technical change was already contained in Ricardo's work (1817); it was subsequently endorsed by Marx (1867), and reasserted later by Hicks (1932). Already Ricardo did note the constant competition between machinery and labor, and the continual wage increases, which would eventually cause the substitution of capital (in the sense of 'better machines') for labor.

Hayami and Ruttan illustrate the validity of their model using the case of Japan and of the USA between 1880 and 1960. These two countries have indeed often been presented as two alternative agricultural models of development. They are characterized by extremely different factor endowments, and consequently by very diverse factor price ratios. The authors point out that in 1960, the 'arable land area per male worker was forty-seven times greater in the United States than in Japan' (Hayami and Ruttan, 1970: 1116), and that in the same year, 'a Japanese farm worker would have to work thirty times as many days as a U.S. farm worker in order to buy a hectare of arable land' (op. cit. 1117). Clearly, the scarcity of land in Japan made its relative price increase sharply over the period, whereas in the USA, the 1880–1920 period witnessed a sharp increase in the price of labor relative to the price of land. The authors note also that, despite these large differences, both countries experienced rapid growth rates of output per worker during the period under review. This was fuelled by an increasing demand for agricultural products in both countries. After a failed attempt in the 1870s to introduce in its agricultural sector machinery and agricultural tools imported from the West, the Meiji government of Japan turned instead to biotechnology. From the following decade, the decrease in the price of imported fertilizers enabled the selection of high-yield varieties, which were more responsive to these new inputs. Consequently, the fall of fertilizer prices in Japan relative to land, and the fall of machinery prices (tractors) relative to wages in the USA, induced an adjustment in factor proportions in both countries. The technologies introduced in Japan tended to save land and to use labor more intensively, whereas in the USA, labor saving technologies were favored.

An illustration of Hayami and Ruttan's model is provided in Figure 5.2. The figure shows the innovation possibility curve I_0 which is the envelope of all possible isoquants, such as i_0 that make production technologically efficient. The innovation possibility curve I_0 shows the range of all possible technologies in time period 0. Production occurs at point A, with L_0 units of labor and N_0 units of land. This is the least cost combination of resources, given the price ratio P_0. If, as in the case of Japan, labor becomes more abundant relative to the other factor in time period 1, the new price ratio becomes P_1. In Figure 5.2, more expensive land prices, relative to the other factors will lead to factor substitution; less land will be used and the use of the labor factor will be intensified. Technical change will set in which will take the form of land-saving techniques. New technologies will be developed, so as to save scarce resources and use abundant resources. Farmers will now adopt the new least cost technologies, which are represented by the new innovation possibility curve I_1, envelope of all i_1 curves. Production will now occur at point C, employing L_1 units of labor and N_1 units of land. Note that point B on isoquant i'_0 represents the least cost combination of factors, with *no* technical change, as we remain on the same innovation possibility curve.

In subsequent work, Hayami and Ruttan (1985) claim that the induced innovation hypothesis can be tested successfully in the case of countries such as Taiwan, Korea, the Philippines, but also in Denmark, France, Germany and the UK. Reviewing American agricultural development since the 19th century, Olmstead and Rhode (1993) criticize Hayami and Ruttan's model for failing on several accounts, such as

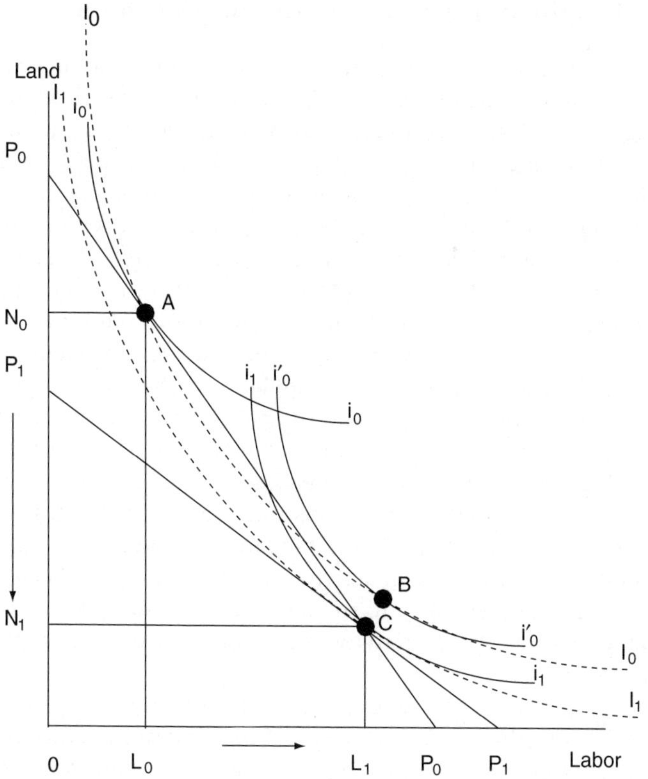

Figure 5.2 A model of induced technical innovation
Source: Hayami and Ruttan (1985).

for ignoring the fact that there were substantial regional differences in the movement of land/labor ratio. Even if important historical 'stilized' facts were overlooked by Hayami and Ruttan, their model is nevertheless useful in shedding light on some aspects of agricultural development in various countries. Moreover, Hayami and Ruttan (1970) highlight the fact that technological change inducement in those countries that have had relatively successful agricultural sectors, was very much conditional on substantial national efforts in agricultural research and extension services. For example, a system of itinerant agricultural instructors was put in place in Japan as early as 1885. The increase in farmers' level of general and technical knowledge is a prerequisite for successful introductions of innovations in the agricultural sector, an issue that is explored in the next section.

5.3 The importance of knowledge in agriculture

Following developments in the field of innovation economics, the idea that successful innovations flourish best in a specific cultural, institutional, societal and economic environment has imposed itself as a dominant paradigm in recent years. Friedrich List's concept of System of Innovation (List, 1841), popularized in the late 1980s by authors such as Nelson (1993) can easily be transposed to the agricultural sector. The Agricultural System of Innovation, referred to also as the Agricultural Knowledge System (AKS) (Blum, 1994), has the following components: farms, governmental organizations and institutions, agricultural research centers and agricultural schools, cooperatives and commercial entities, including the suppliers of agricultural inputs. The key component of the AKS is the *transfer of knowledge* from research centers, from governmental institutions or from other farms that lie at the cutting edge of the technology frontier, to all farms in a given country. Governmental organizations and research institutions provide the necessary extension services to the farming community. Extension services can be defined as all activities undertaken by advisory bodies to diffuse new knowledge into the farming community so as to affect favorably its level of productivity and income, without damaging the environment.

There is a plethora of studies dealing with the importance of knowledge and education to the successful absorption of innovations in all sectors of the economy. For example, Hoag *et al.* (1999) show how the educational level, as well as other characteristics such as experience, influence the rate of adoption of computers in US farms in 1995. Cioffi and Gorgitano (1998) find that the local innovative capability, or local *milieu* with its network effects, has a definite impact on the adoption of computers by Italian farmers. The authors call for a broader role of extension services in the localities with lower rates of innovation.

The AKSs help identify specific agricultural problems, and to disseminate existing scientific knowledge so as to develop a technological solution to the problems. It is quite easy to determine and to analyze the specific AKSs at each national level. For example, the analysis by Rogers *et al.* (1976) attempted at identifying the key factors that made the US Co-operative Extension Service successful over time. The authors came up with eight such factors explaining the success of the US AKS, such as the orientation of the research sub-system towards utilization, and the availability of a critical mass of technology. The Israeli AKS can also be categorized as one that has been relatively successful. Among the elements contributing greatly to this success, one can cite (Blum, 1991): a good educational level of the knowledge users, and an important role played by the media and farmers in the exchange of agricultural knowledge. An additional positive element of geographically small AKSs is the easiness with which network effects can take place. These network effects are explained by sociological and cultural forces, that characterize small territorial economic units such as the region. Being one such small country in Europe, The Netherlands boasts nevertheless high agricultural output ratios. A brief analysis of the factors lying behind the Dutch success story is suggested in Box 2.

Box 2 Explaining the supremacy of the Dutch agricultural system

The Dutch have been, for quite some time, the technical and trade leaders in European agriculture. Their yields are among the highest and their agricultural incomes per capita are among the highest in the EU. If one includes their trade with the other EU partners, The Netherlands is, despite their small size, one of the world's largest agricultural exporters. Their exports are concentrated mostly on high value output, with a heavy specialization in horticulture, dairy products and cereal-based livestock products. In particular, horticulture has been facilitated by the use of green houses and has enabled the Dutch farmers to produce flowers, lettuce, cucumbers and tomatoes, mostly for the north European market in spite of the availability of cheaper products originating from Italy and from other south European countries. Although Dutch selling prices for these products on North European markets have constantly been above those of south European products, the sophisticated and advanced organization of the Dutch agribusiness system has meant that Dutch agricultural products have been able to withstand the competition from southern producers. This successful export penetration has meant that Holland has acquired the reputation of the EU country *par excellence* of farmers/entrepreneurs.

It is clear that the development of an industrial and mercantile agricultural system over many decades explains today's success, for the commercial and technical superiority of Dutch farmers is a direct result of an appropriate AKS. In particular, the Commodity Boards (*Produkt Schappen*) and the agricultural research institutions

Box 2 (Continued)

represent the two major pillars of the Dutch AKS. The vertical structure of the Commodity Boards enables it to cover all stages in the production, processing and distribution of food products. Vertical integration favors in turn many synergies between farms and the food industry, and is conducive to cost minimization strategies. As shown by Haveman (1954), what enabled a radical change in the farmer's economic role and economic standing in the late 18th century north-east Dutch society, was his attitude towards new ideas, towards theoretical research, in short, towards knowledge. The new serum discovered towards the end of the eighteenth century to fight the cattle plague that had devastated many herds in previous years, was proof that scientific discoveries could be used to master the powers to which farmers had always been subjected. From then onwards, it became clear to the Dutch farmer that greater prosperity could be achieved continuously through the absorption of new ideas in the field of agricultural science. It is from that time that the farmer in Holland started to behave as a true *homo economicus*, taking risks, producing with the objective of generating high profits, and competing in a new individualistic societal environment. His faith in scientific progress was such that the farmer entrepreneur would send his sons to study at the University of Groningen, sowing the seeds of many generations of well-educated farmers in the country.

Moreover, the introduction of new ideas in the field of agricultural production, coupled with an intensification of production methods, led to a substantial competitive advantage allowing Dutch agricultural products to penetrate new markets beyond the boundaries of the farmer's village, and even more so, to break into foreign markets.

5.4 Agricultural innovation in the EU

This section will present and analyze agricultural innovation in the EU by focusing first on the input indicators of innovation, such as the funding issues, primarily at the EU level. It will then conclude with a discussion on the returns to agricultural research and innovation, both private and public.

5.4.1 Input indicators: research funding and extension services

5.4.1.1 Community funding

Among the Community policy instruments enabling the gestation, development and transfer of agricultural innovation, one can refer to the Guidance section of the European agricultural fund (EAGGF), as well as to the research programs specifically targeted at agriculture, but also at fisheries and forestry sectors. As was seen in Chapter 4, the Guidance section of the EAGGF is aimed at promoting structural change in agricultural holdings, through the introduction of modern farming techniques. During the 2002 financial year, the portion of the EU budget dedicated to the Guarantee section of the EAGGF amounted to €2639 million; this represented 5.9 per cent of the EAGGF, and a bare 2.7 per cent of the total commitment appropriations for the year (CEC, 2001b).

The research and technological programs specifically designed for the agricultural sector are embraced in what is commonly referred to as the Community Multi-annual Framework Programs (MFP). MFP I ran from 1984 to 1987, with an overall budget of ECU 3250 million. Subsequent programs were: MFP II (1987–91) with ECU 5396 million, MFP III (1991–94) with ECU 6600 million, MFP IV (1994–98), with ECU 13,215 million and MFP V (1998–2002) with ECU 14,960 million. The aim of the framework programs has been and still is to foster scientific and technical excellence so as to improve the competitiveness, economic growth, employment prospects and social cohesion in Europe. The theme of 'economic and social cohesion' was introduced in the Treaties through the 1986 Single European Act (SEA); it is also the SEA that made science a Community responsibility. The various MFPs are carried out in line with the objective set out in article 163(1) of the Treaty establishing the European Community, which is that:

> of strengthening the scientific and technological bases of Community industry and encouraging it to become more competitive at international level, while promoting all research activities deemed necessary by virtue of other Chapters of this Treaty (Treaty of Amsterdam, Title XVIII).

Articles 164 to 173 determine the activities to be carried out within the scope of the MFP. Although article 81 of the EU competition policy (formerly article 85 in the Rome Treaty) prohibits every agreement or strategy between firms that distort appreciably the level of competition in the EU, R&D collaboration is one of the exceptions envisaged in this article.

The restriction of collusive behavior does not therefore apply to the case of technological and research collaboration in the EU, particularly at its pre-competitive stage. Moreover, technological collaboration takes the form of selective financial support for research projects that involve firms and research institutions from various countries of the EU. The financial support channeled through the EU budget is conditional on co-financing from the member states. Although the ceiling on resources to be pledged for research and technological development represented slightly more than 4 per cent of the EU total during the 2002 financial year, it should be noted that the disproportionate EU budgetary spending on the Guarantee section of the EAGGF for many decades left little room for other policies until recently. The February 1988 Council Decision on the introduction of a more stringent financial discipline in order to put an end to declining budgetary resources and to uncontrolled CAP expenditure, led to an agreement between Council of Ministers, Commission and Parliament in June 1988 (see Chapter 4). This agreement took the form of the Delors I package, and enabled, *inter alia,* the doubling of the structural funds, in absolute terms, including the Guidance section of the EAGGF. It is from this time that the other policies, such as the research and technological policy assumed a greater role.

Although the manufacturing sector, and in particular the information technology-related industries were given priority from the onset, agriculture has also been one of the broad sectors targeted by the Commission. Consequently, it has featured quite prominently in the list of research programs. In the Commission's Second Framework Program for research (1986–90), there were five sub-programs dealing with agriculture, fisheries and forestry alone. These were: FLAIR, which promoted research in food science and technology; ÉCLAIR, which was the first programme of its kind on bio-technology-based agro-industrial research and technological development; CAMAR, a research and technological development program dealing with the competitiveness of agriculture and the management of agricultural resources (Table 5.5). To these programs, one can also add: FAR and FOREST (Forestry Sectoral Research and Technology), the latter covering raw materials, recycling, and renewable materials in the area of forestry and wood products. The Commission's Third Framework Programme (1991–94) added AIR (Agro-Industrial Research) to the list. Research under this program was organized into four scientific areas, such as: primary production in agriculture, horticulture, forestry, fisheries and aquaculture; inputs to agriculture, forestry, fisheries and agriculture; processing of biological raw materials; and end use products.

As can be seen from Table 5.5, biotechnology is a central theme in EU-sponsored collaborative programs for agriculture. Many programs such as ÉCLAIR, BRIDGE, BAP and BIOTECH were aimed at improving the basic knowledge of living systems and at increasing productivity through applications to agriculture, health, nutrition, the environment and industry. In particular, ÉCLAIR aimed at encouraging the production and testing of new products derived from agricultural raw materials, as well as the development of new inputs such as fertilizers, pesticides, vaccines, inoculants and growth promoters.

The structure of the life science/biotechnology industry in Europe and in the world is characterized by the juxtaposition of a few dominant firms with a large number of small SMEs. The group of large firms in Europe includes BASF, Bayer, AgrEvo, Rhône-Poulenc, as well as the Swiss Novartis Crop Protection. In 2001, there were more than 1500

Table 5.5 A synopsis of the main European programs promoting biotechnology research

Acronym and full name	Period	Budget (ECU mio)	Prime objectives
ÉCLAIR: European Collaborative Linkage of Agriculture and Industry through Research	1988–93	80	Applying advanced biotechnology in agro-industrial sector, using especially raw materials from agriculture
FLAIR: Food Linked Agro-Industrial Research	1988–93	25	As ECLAIR, but food oriented, in manufacture and processing
BAP 1: Biotechnology Action Program	1985–89	75	Develop infrastructure in biotechnology, especially research and training
BAP 2: Biotechnology Action Program	1990–93	100	As BAP 1, but for large projects, e.g. Molecular modeling, advanced cell culture
BRIDGE (Biotechnology Research for Innovation, Development and Growth in Europe)	1990–93	100	Developments in the area of biotechnology
BIOTECH	1992–96	164	Pre-normative research, more basic than BRIDGE, includes safety
FAIR: Field of Research Agriculture Forestry and Rural Development	1994–98	15	Research in the area of transmissible spongiform encephalopathy

Source: EC Commission Annual Report on Research and Technology Development, DGXII.

biotechnology companies in the EU and Switzerland, employing roughly 61,000 people. This compares with 1273 firms and 162,000 employees in the USA (CEC, 2002a). Consequently, European biotechnology firms are typically small, and much smaller than their US counterparts. Clearly, the larger average size of US firms confers indisputable advantages to American companies over their European counterparts, particularly in an industry where survival and growth depend on innovation, and where research costs are often prohibitive.[8] As a result, the difference in performance between US and European firms appears clearly in terms of their revenue: US firms generate three times as much revenue as the EU and Swiss firms. As for the research profile of these firms, a number of mergers and acquisitions since the mid-1990s have led to the concentration of biotechnology research in the hands of a few major companies such as AgrEvo, DuPont de Nemours and Monsanto, both US firms. The strategy of these firms has been to integrate vertically: (i) upstream by entering agreements with genomics companies, so as to increase their research capability; and (ii) downstream by investing further down in the food chain.

However, given the hostility that engineered food production has met in Europe, particularly in continental Europe, the EU is only marginally involved in the production of GMOs, when compared with the USA, and its agro-food activities have actually declined sharply in recent years. The European firms are indeed more active in the areas of genomics, combinatorial chemistry and bioinformatics (CEC, 2002b).

It should be noted that, not surprisingly, the products that are subject to genetic manipulations are for the most part crops of economic importance for western countries (in particular for the USA), in terms of both production and trade. In the year 2000, the total world area under GMOs was broadly divided as follows: 58 per cent for soybeans, 23 per cent for maize, 12 per cent for cotton and 7 per cent for canola (James, 2000). In the same year, the USA had the lion's share of the total area under GMOs (68 per cent), whereas the share of European countries remains minuscule, with Spain, France and Romania representing each less than 1 per cent of the world area.

The aim of all the EU programs has been to simulate technological collaboration across EU countries, so as to avoid the costly and inefficient duplication of research efforts. These joint efforts, co-funded through the EC budget were designed ultimately to make agricultural and fisheries holdings more competitive, to improve rural and forestry management and to foster health and environment protection in the area of fisheries and aquaculture. At this point, it is pertinent to ask whether EU funding was worth spending in this way.

Any information on the evaluation of past collaborative programs is normally difficult to obtain. One of the exceptions is ÉCLAIR, the evaluation of which was undertaken in 1999. Some 42 projects co-funded under ÉCLAIR and representing a budgetary outlay of around €65 million were assessed. These 42 projects were pre-competitive in nature, implying that further research, and therefore further funding is necessary to bring the projects closer to the commercialization stage. The selected projects were all aimed at developing new inputs for the agricultural sector such as fertilizers, pesticides, inoculants, vaccines and growth promoters. It was found that approximately a third of all 42 projects had been successful in commercializing a product six years after EU funding had ceased; another third had reached the prototype or trial stage, whereas 14 per cent of the projects had developed products that were to be commercialized in the near future. The nature of the different products ranged from the production of a new vaccine preventing diseases in fish, to potatoes and other crops with modified characteristics.

In recent years, the focus of EU research and technological collaboration has shifted to supporting more sustainable agricultural practices and to promoting food safety. Indeed, the outbreaks of BSE and foot and mouth diseases, the vivacious debates in Europe on the ethical and socio-economic implications of biotechnology and GMOs, as well as the acknowledgement that past agricultural success has had a detrimental effect on parts of the European environment and rural heritage, led to a new orientation in EU policy thinking. The EC recognized these problems by including support for more sustainable agricultural practices in the 1997 Amsterdam Treaty. This new way of thinking has promoted collaborative research in the area of rural development, with the objective of defining proper tools of management for landscapes and ecosystems that support biodiversity. Collaborative research is also encouraged in the area of more competitive organic farming methods, and in the use of genetics to improve the safety and quality of animal systems. One concrete outcome of this new direction is the recent creation of a European Food Safety Authority (OJ L, 2002). In particular, although the crops grown with organic farming methods represented only 3 per cent of the total EU utilized agricultural area in 2000, the organic farming sector has grown by more than 25 per cent every year since 1993.

The sixth Multi-annual Framework Program (MFP VI) covers the four-year period 2002–06, with a budget of €17.5 billion. This represents an increase of 17 per cent compared with MFP V, which is encouraging, given the smaller and controversial increase that the fifth framework program was able to obtain. Although the main innovative focus of MFP VI is the idea of creating 'a European research area', through a symbiosis between the research policy proper and the other community policies, the dual emphasis on food quality and safety, and on sustainable development is reiterated fully. Food safety and sustainable development are two among seven thematic priorities contained in the new program. Here again, biotechnology tools are seen as being essential for the conduct of research on food quality and safety. As a consequence, the field of life sciences is given first priority, as it is seen to respond adequately to the objective of the European Community's Lisbon Summit, which is that of allowing the EU to become a leading knowledge-based economy within a decade or so. MFP VI is seen by the Commission as one of the tools to help restore European leadership in life sciences and biotechnology research. The emphasis of the co-funded research programs in this area will be on the whole food chain, from the farm to the final consumer. This involves in particular activities in terms of 'traceability', with a specific emphasis on GMOs. Traceability refers to the ability to prove the origin of food products. This concept has already been introduced in EC law by the July 2000 EC regulation N.1760/2000 of the European Parliament and the Council of Ministers. Specific labeling is now required in the case of beef and of other beef products, so as to allow the transparency of information to the benefit of consumers. The label needs to specify clearly the geographical origin of the animal, and of all the intermediaries in the case of processed meat. This regulation has entered into force in August 2000, and is directly applicable to all EU member states. More emphasis on labeling and traceability will eventually enable European consumers to differentiate clearly between genetically modified and non-genetically modified products in the future.

Finally, in addition to sustainability of farming methods and to food safety, another new orientation of EU agricultural research has emerged during the 1990s. Through the creation of EIARD (European Initiative for Agricultural Research for Development), a new scheme of cooperation between European countries and the developing world has been implemented since 1995 (CEC, 1997). The emphasis of this scheme is on collaborative research in all areas of agriculture with the purpose of helping developing countries. The objective of EIARD is to improve the impact of investments in R&D by a more efficient coordination between its 18 partners (that is EU-15, Norway, Switzerland and the Commission). Europe is currently still involved in current discussions to build a Global Forum of Agricultural Research for Development. This would bring together all the actors in agricultural research worldwide both at policy and institutional level. Regional fora exist and have been held for example, in sub-Saharan Africa in 1996. Representatives of the different regional fora met for the first time in May 1996, and the first Global Forum meeting took place in October 1996. This complements in some way the work of two European Agricultural Research Organizations (NATURA and ECART).[9]

5.4.1.2 Funding and extension services at the national level

It should be borne in mind that the increase in EU collaborative research since the 1980s nevertheless still leaves most research at national level. Most of the research effort in EU countries is actually funded by national budgets and most, if not all, extension services agencies are nationally based. The number of research institutions dealing with agriculture in the EU is quite large, and their type is quite diverse. All we can suggest is therefore a broad overview of national agricultural research and extension systems, concentrating only on the most important EU producing countries, that is Holland, France, Italy and Ireland.

The Netherlands

The Dutch AKS encompasses nine major institutes for Applied Research, several agricultural research centers, as well as various organizations dealing with extension, education and training; these different constituent parts of the AKS keep close contacts with one another. Each institute for Applied Research is specialized in a given research area, such as for example in the case of the Research Institute for Pig Husbandry. These institutes conduct research with a view to allowing direct applications by farmers and crop growers. Apart from the centrality of innovation in Dutch agriculture, another distinguishing factor of the Dutch AKS is its extensive and impressive ramification in the EU and in the rest of the world. There are two major channels that have permitted the dissemination of Dutch agricultural know-how and presence in the world: one is through cooperation in agriculture; the other is through advisory services beyond national borders.

First, the IAC (International Agricultural Center) facilitates co-operation in agriculture, the environment, rural development, nutrition and fisheries with developing and with Central and East European countries. The Center offers training courses for professionals abroad, and working in close collaboration with the several research centers at Wageningen

University, it provides logistic support for the dissemination of research results through international conferences in Holland. Second, the Agricultural Research Institute (Landbouw Natuurbeheer en Visserij) is also an important component in terms of the internationalization of Dutch agriculture, nature management and fisheries. Employees of the institute work on 40 locations all over the world to assist Dutch Agricultural SMEs (Small and Medium-Sized Enterprises) in their activities in foreign markets. These are mostly scattered in Europe, including Russia, but they extend as far as south-east Asia, South America and eastern Africa. As discussed by van der Eng (1996), Dutch research expertise in the country's ex-colonies was such that by the beginning of the Second World War, most agricultural research institutes in Bogor (Indonesia) had acquired an international reputation in the area of tropical agriculture.

France

The French *INRA* (Institut National de la Recherche Agronomique) is one such key institution providing research and extension services to the French farming and agribusiness sectors. Created in 1946 as a public institution, the *INRA* is linked to the two French Ministries of Research and Agriculture. Its main objective is to produce and to disseminate research findings and innovations in the areas of agriculture, food and the environment. Its annual budget of €573 million allows the funding of an efficient organizational structure of more than 250 research units, 80 laboratories and more than 130 advisory units disseminated all over the French territory. The *INRA* has a large ramification of collaborative links, not only with EU research centers but also with research units in other parts of the world.

Italy

There are a number of bodies dealing with agricultural research and extension services in Italy. A prominent one is the *IRSA* (Istituto di Ricerca e Sperimentazione Agraria), which is part of the Ministry for Agricultural and Forestry Policy, and which covers many fields of research from agronomy, in particular irrigation, to eonology (wine production). The extension services provided by *IRSA* enable for example the increase in computer literacy of the Italian farming community, with possibilities of co-funding for the acquisition of hardware (see Law of 15 November 2001). Beside the *IRSA* lie a number of other institutes embraced by the *CNR* (Consiglio Nazionale delle Ricerche), such as for example, the Center for the Study of Agricultural and Rural Systems

Management (Centro di studi sulla Gestione dei Sistemi Agricoli e Territoriali), linked to Bologna University.

Ireland
The Irish institution *TEAGASC* (Irish Agriculture and Food Development) provides training, advisory as well as research services. On the training front, it offers third level and vocational courses to young people planning to enter careers in the agricultural, horticultural and agri-food sectors. *TEAGASC* has contacts with some 80,000 farmers and rural dwellers, and helps them apply the results and findings of the institution's research programs.

5.4.2 The rates of return to agricultural research and innovation

The rates of return to agricultural research and innovation must be appraised at both the private and public (social) level. There may be indeed social costs generated by innovation in the agricultural sector, such as those arising from the use of pesticides, given their possible detrimental impact on the environment and on human health. A comprehensive cost-benefit analysis encompassing private as well as social dimensions is very difficult to undertake, and this is why most of the literature has concentrated so far on some partial aspects of the problem. For example, Esposti (2000) attempts at calculating the internal rate of return of investment in R&D in Italian agricultural holdings over the period 1971–95, and concludes that R&D has been effective within a few years of investment, particularly on the output side (animal products). In his review of studies on the measurement of the returns to human capital investment, Evenson (1990) finds that most studies find a high return to agricultural research, thereby justifying additional national spending in research and extension programs.

Summary

This chapter has highlighted the importance of knowledge and of innovation to economic growth in general, and to the agricultural sector in particular, from both a theoretical and applied angle.

It has shown how the introduction of new knowledge and techniques in the agricultural sector has enabled the increase in food production at a rate faster than that of the population growth, particularly since the industrial revolution, proving Malthus' predictions wrong. Part of food and agricultural research is undertaken in a collaborative manner at the EU level, although most of the research is conducted at the national level. Biotechnology is one central element of modern agricultural research in developed as well as developing economies. In Europe, the dominant perception is that the intensification of agricultural production through genetic engineer should not be done at the expense of food safety. The Treaty of Amsterdam has reiterated the new orientations of EU research policy for the agricultural sector in the 1990s: sustainable development, food safety and research collaboration to the benefit of developing nations are three important elements of this policy. The European Community supports a coherent, comprehensive, effective and transparent approach to biotechnology across international fora such as the FAO and the WTO.

Key terms and concepts

Endogenous growth
Structural change
Induced innovation
Extension services
Biorevolution
Genetic engineering
Knowledge transfer and knowledge spillovers
Sustainable development

Notes

1 Note that in the standard production theory, the marginal product of inputs declines, implying that sustained growth is only possible through exogenous technological change (see Chapter 2).
2 It should be noted that the use of key concepts elaborated by modern growth theorists has not been confined to neoclassical based models. Laibman (1981) provides one such example of an endogenous technical change model based on Marxian conceptual foundations, such as the split of the production sphere into two sectors: the capital goods and the consumer goods sectors.
3 For a very thorough review on innovations over time, see in particular Mokyr (1990).
4 Sustainable development, a popular concept since the mid-1980s, can be defined as 'development which meets the needs of the present without compromising the ability of future generations to meet their own needs' (WCED, 1987).
5 These firms were: Novartis, Monsanto, Du Pont de Nemours, Zeneca, AgrEvo, Rhône Poulenc, Bayer, American Cyanamid, Dow Agrosciences and BASF.
6 In the light of this uncertainty, it is worth inviting the scientific community involved in certain aspects of biotechnology research to meditate upon Rabelais' famous maxim: '*Science sans conscience n'est que ruine de l'âme*'.

7 See for example the cross-industry study carried out in Mansfield's work (Mansfield, 1968).
8 On the advantages conferred by size, see for example Jacobson and Andreosso-O'Callaghan (1996).
9 NATURA (Network of European Agricultural Tropically and sub-tropically oriented Universities and scientific complexes Related with Agricultural development). ECART (European Consortium for Agricultural Research in the Tropics).

Bibliography

Abramovitz, M. (1956) 'Resource and output trends in the United States since 1870', *The American Economic Review,* 46: 2, May, 5–23.

ADB (2001) *Agricultural Biotechnology, Poverty Reduction, and Food Security.* Asian Development Bank, Manilla.

Anderson and Herdt (1988) 'The impact of new technology on foodgrain productivity to the next century', in *Agriculture and Government in an Interdependent World,* Proceedings of the Twentieth International Conference of Agricultural Economists, 683–96.

Arrow, K.J. (1962) 'The economic implications of learning by doing', *Review of Economic Studies*, 29, June, 155–73.

Babbage, C. (1832) *On the Economy of Machinery and Manufactures* (London: Knight), reprinted (1993) London: Routledge/Thoemmes Press.

Blum, A. (1991) 'What can be learned from a comparison of two agricultural knowledge systems? The case of The Netherlands and Israel', *Agriculture, Ecosystems and Environment,* 33, 325–39.

Blum, A. (1994) 'A Confederate agricultural knowledge system: the special case of Switzerland', *European Journal of Agricultural Education and Extension,* 1: 1, April, 103–17.

CEC (1997) *European Initiative for Agricultural Research for Development – EIARD,* COM (97) 126 final, Brussels.

CEC (2001a) *Economic Impacts of Genetically Modified Crops on the Agri-Food Sector – A First Review,* DG for Agriculture, Brussels.

CEC (2001b) *Budget Général de l'Union Européenne pour l'Exercice 2002,* Bruxelles, Luxembourg, Janvier.

CEC (2002a) *Life Sciences and Biotechnology – A Strategy for Europe,* COM (2002) 27 final, Brussels.

CEC (2002b) *Innovation and Competitiveness in European Biotechnology,* Enterprise Directorate General, Brussels.

Cioffi, A. and Gorgitano, M.T. (1998) 'Local innovative capability and innovation adoption in agriculture: computers on Italian farms', *European Review of Agricultural Economics*, 25: 2, 210–27.

Cohen, W.M. and Levinthal, D.A. (1990) 'Absorptive capacity: a new perspective on learning and innovation', *Administrative Science Quarterly*, 35, 128–52.

Coombs, R., Saviotti, P. and Walsh, V. (1987) *Economic and Technological Change* (London: Macmillan).

Ellis, W. (1826) 'Effect of the employment of machinery upon the happiness of the working classes', *Westminster Review,* 5, January, Article IV.

Esposti, R. (2000) 'The impact of public R&D and extension expenditure on Italian agriculture: An application of a mixed parametric/nonparametric approach', *European Review of Agricultural Economics,* 27: 3, 365–84.

Evenson, R.E. (1990) 'Human capital and agricultural productivity change', in *Agriculture and Governance in an Interdependent World,* Proceedings of the Twentieth International Conference of Agricultural Economists (Aldershot: Dartmouth).

Fabricant, S. (1954) *Economic Progress and Economic Change,* 34th Annual Report of the National Bureau of Economic Research (New York).

Gaffard, J.-L. (1990) *Economie Industrielle et de l'Innovation* (Paris: Précis Dalloz).

Griliches, Z. (1957) 'Hybrid corn: an exploration in the economics of technological change', *Econometrica,* 25: 4, 501–22.

Haveman, J. (1954) 'Social tensions between farmers and farm laborers in Northern Holland', *American Journal of Sociology,* 60: 3, November, 246–54.

Hayami, Y. and Ruttan, V.W. (1970) 'Factor prices and technical change in agricultural development: The United-State and Japan, 1880–1960', *The Journal of Political Economy,* 28: 5, September/October, 1115–41.

Hayami, Y. and Ruttan, V.W. (1985) *Agricultural Development: An International Perspective* (Baltimore: Johns Hopkins University Press).

Heertje, A. (1977) *Economics & technical Change* (London: Weidenfeld and Nicolson; First published in Dutch in 1973, by H.E. Stenfert Kroese B.V.).

Hicks, J.R. (1932) *Theory of Wages* (London: Mac Millan).

Hoag, Dana L., Ascough, James C. and Fraiser, W. Marshall (1999) 'Farm computer adoption in the great plains', *Journal of Agricultural and Applied Economics,* 31: 1, April.

Jacobson, D. and Andreosso-O'Callaghan, B. (1996) *Industrial Economics and Organization – A European Perspective* (London: McGraw&Hill).

James, C. (2000) *Global Review of Commercialized Transgenic Crops in 2000.* ISAAA Brief No. 21. Ithaca, New York: International Service for the Acquisition of Agribiotech Applications.

Kuznets, S. (1930) *Secular Movements in Production and Prices* (Boston and New York: Houghton Mifflin).

Laibman, D. (1981) 'Two-sector growth with endogenous technical change: a Marxian simulation model', *The Quarterly Journal of Economics,* 96: 1, February, 47–75.

List, F. (1841) *National System of Political Economy* (London: Dent).

Lucas, R.E. (1988) 'On the mechanics of economic development', *Journal of Monetary Economics,* Vol. XXII, 3–42.

Maddison, A. (2001) *The World Economy – A Millennial Perspective* (Paris: OECD, Development Centre).

Malthus, T.R. (1798) *An Essay on the Principle of Population* (London: New Penguin Classics Edition, 1988).

Mansfield, E. (1968) *The Economics of Technical Change* (New York: Norton).

Marx, Karl (1867) *Das Kapital* (Reprinted in English by J.M. Dent, London).

Mokyr, J. (1990) *The Lever of Riches – Technological Creativity and Economic Progress* (New York and Oxford: Oxford University Press).

Nelson, R.R. (ed.) (1993) *National Innovation Systems: A Comparative Analysis* (Oxford: Oxford University Press).

OJ L (2002) *Official Journal of the EC* (OJ L 31 of 1.2.2002).

Olmstead, A. and Rhode, P. (1993) 'Induced innovation in American agriculture: a reconsideration', *Journal of Political Economy,* 101: 1, 100–18.

Phelps, E.S. (1966) 'Models of technical progress and the golden rule of research', *Review of Economic Studies,* 33, April, 133–45.

Ricardo, D. (1817) *On the Principles of Political Economy and Taxation.* Reprinted in P. Sraffa and M. Dobb (eds) (1953) *The Works and Correspondence of David Ricardo,* Vol. I (Cambridge: Cambridge University Press).

Rickson, R.P., Saffigna, F., Vanclay, F. and McTainsh, G. (1987) 'Social bases of farmers' responses to land degradation', In: A. Chisholm and R. Dumsday (eds) *Land Degradation, Problems and Policies* (Cambridge: CUP).

Rogers, E. (1983) *Diffusion of Innovations* (New York: Free Press).

Rogers, E.M., Eveland, J.D. and Bean, A.S. (1976) *Extending the Agricultural Extension Model,* Institute for Communication Research. Standford University, CA.

Romer, P.M. (1986) 'Increasing returns and long-run growth', *Journal of Political Economy,* 94: 5, 1002–37.

Santikarn Kaosa-ard, M. and Rerkasem, B. (2000) *The Growth and Sustainability of Agriculture in Asia* (Oxford, New York: Oxford University Press).

Schumpeter, J.A. (1912) *Theorie der Wirtschaftlichen Entwicklung* (Leipzig: Duncker und Humboldt).

Schumpeter, J.A. (1943) *Capitalism, Socialism and Democracy* (London: Allen and Unwin).

Simpson, J. (1995) *Spanish Agriculture – The long siesta,* 1765–1965 (Cambridge: Cambridge University Press).

Solow, R.M. (1957) 'Technical change and the aggregate production function', *The Review of Economics and Statistics,* 39: 3, 312–20.

Uzawa, H. (1965) 'Optimum technical change in an aggregative model of economic growth', *International Economic Review,* 6, January, 18–31.

Vanclay, F. and Lawrence, G. (1994) 'Farmer rationality and the adoption of environmentally sound practices: a critique of the assumption of traditional agricultural extension', *European Journal of Agricultural Education and Extension,* 1: 1, 59–90.

Van der Eng, P. (1996) *Agricultural Growth in Indonesia – Productivity Change and Policy Impact since 1880* (London and New York: Macmillan and St. Martin's Press).

Walras, L. (1874) *Eléments d'Economie Politique Pure ou Théorie de la Richesse Sociale* (reprinted 1952; Paris: Librairie Générale de Droit et de Jurisprudence).

Watson, A. (1983) *Agricultural Innovation in the Early Islamic World* (Cambridge: Cambridge University Press).

WCED (1987) *Our Common Future* (World Commission on Environment and Development; Oxford: Oxford University Press).

Zilberman, D. (1997) 'The value of economic research', *American Journal of Agricultural Economics,* 79: 5, 1539–44.

6
A Wider European Union: Enlargement to Central and Eastern Europe, Cyprus and Malta

Objectives of this chapter

- To discuss briefly previous enlargements, allowing the then Common Market of six member countries to expand to a European Union of 15 members. In particular, a short discussion of the fourth enlargement and its consequences will be provided so as to introduce a comparative framework.
- To analyze the main opportunities and challenges presented by the fifth enlargement to both the Central and East European Countries (CEECs) and to the current EU.

Introductory remarks

The enlargement of the EU to include ten CEECs is the fifth of its kind since the Common Market was created in 1958.

The first enlargement in 1973 allowed the expansion of the then Common Market to Denmark, Ireland and the United Kingdom. Membership of a large country such as the UK, with a relatively smaller and more efficient agricultural sector than that of other Common Market members at the time, induced the need for the first substantial review of the common budget. Being a net importer of foodstuffs with smaller agricultural spending requirements, the UK was clearly going to be a net loser in budgetary terms. This led to the introduction of the European Regional and Development Fund (ERDF), which at least was beneficial to some poorer regions of the UK, such as Northern Ireland. After the enlargement to Greece in 1980, the main attention of EU policy makers shifted to the likely risks of overproduction in many sectors, brought about by the ensuing enlargement to Spain and Portugal in 1986. Wine, vegetables, fruits and olive oil were the sectors most affected by the risk of over production at the time. Accommodating the new Iberian peninsula members in the then European Community led to a substantial reform in these sectors, with, for example, the introduction of an allowance for the eradication of poor quality vineyards. The inclusion of ex-EFTA members (European Free Trade Area) in the EU in 1995, was, from the Union's point of view, the smoothest of all enlargements experienced. In particular, Austria, Finland and Sweden were already enjoying free trade links with the EU with the creation in the early 1990s of the European Economic Area (EEA) between the European Community and the EFTA.[1]

Whereas the various enlargements so far have generally been gradual and have incorporated countries with market-based economic systems, the fifth enlargement is significantly different in that it represents, for most applicant countries, the culminating point of a transition process linking two radically different economic systems. Although the countries from Central and Eastern Europe have signed a number of Europe Agreements with the EU, preparing them gradually to full membership, their inclusion in the EU in 2004 has major implications, first on the financing of the EU, and in particular of the CAP, second on their domestic economic system, and in particular on their agricultural system.

6.1 Previous enlargements

In some ways, the difficulty in integrating new member countries with substantially lower levels of wealth and with much bigger and relatively inefficient agricultural sectors had already been experienced by the EC in 1986 when it enlarged to Spain and Portugal. The lengthy negotiations were aimed at trying to resolve the many problems felt at the level of the agricultural sector. The Spanish and Portuguese agricultural sectors were accorded a transition period of ten years, during which the common price support mechanisms were progressively introduced. Full membership of CAP, with the ability to avail of EU funding under the EAGGF Guidance section was obtained in subsequent years. For example, Portugal became a full member of CAP on 1 January 1995.

6.1.1 Enlargement to Austria, Finland and Sweden

Shortly after the collapse of the Berlin wall, four of the EFTA member countries applied for full membership to the EU. Austria applied in 1989, Sweden in 1991, Finland in 1992 and Norway in 1992. The negotiations were to be completed by 1 March 1994, and the target date for entry was set on 1 January 1995. Agriculture, fisheries, as well as financial and budgetary provisions featured among the central issues in the negotiations, alongside with regional policy, transport and Community Institutions. Only three countries became members of the EU at the end, as in yet another referendum, Norway rejected membership.

The overall negotiations and the inclusion of these new members in the EU had generally been somewhat facilitated by their belonging to the EFTA and to the EEA prior the enlargement. Before becoming embraced in the EEA, the two regions were economically interdependent, although in an asymmetric way. Indeed in 1990, the EFTA countries represented a quarter of all extra-EC trade; this is to be compared with roughly 18 per cent for the USA. In the same year, the EC market absorbed two thirds of all EFTA exports, and the EFTA countries sourced three quarters of all their imports from the EC market. Agriculture was however not covered by the principle of free movements of goods under the EEA. Fishery products were subject to separate agreements between the EC and Norway on the one hand, and the EC and Iceland on the other. Certain duty concessions were made in return for fishing rights in Nordic waters.

A 'big-bang' approach is what best characterized the inclusion of Austria, Finland and Sweden in the EU on 1 January 1995, in that, in contrast with previous experiences, the three relatively wealthy new members adjusted to EU price levels without any transition period. This posed some socio-economic problems to these three countries for their level of agricultural protection was much more generous during the pre-enlargement period than that of the EU-12. Indeed, farmers from the ex-EFTA countries had to adapt abruptly to the lower EU agricultural prices. Support prices went down in Austria and Finland, whereas in Sweden the fall had been smoothed by the 1992 devaluation of the Swedish krone (see Box 1).

In Austria, farmers were accustomed to a high degree of protection for their domestic production, and to a relatively important financial support for social, regional, ecological and other agricultural functions. Prior accession, the prices for agricultural goods in Austria were 20 to 30 per cent above EU average. Accession to the EU and to the CAP principles meant a reduction of all Austrian agricultural prices to the EU levels, with the exceptions of timber and wine, two sectors where prices increased. Here again, the farmers were compensated for the income loss resulting from lower prices. This income loss had been estimated by the professional association *Oesterreichs Bauern* (1993) at Austrian Schilling 7.8 billion, that is ECU 574.5 million.

Box 1 Finland's accession to the EU

> The experience of Finland's accession to the EU is an interesting case study. Finland is a small open economy, very dependent on trade. Since the disintegration of the Soviet Union, which used to absorb around 20 per cent of its exports, Finland has redirected its trade towards the EU countries. The introduction of CAP instruments led to a sharp decrease in support prices in this country. On 1 January 1995, the new price of barley was 50 per cent that of the December price, and the new price of wheat was 38 per cent the December 1994 producer price. The positive aspect of these price decreases was that consumers were generally able to enjoy cheaper food products. Food prices dropped by 8 to 9 per cent on average shortly after accession. This was reflected in the very low inflation rate, which in September 1995, was, at 0.3 per cent, the lowest in the EU. Understandably, the immediate negative impact on farmers' incomes was softened by means of temporary income transfers. However, the official thinking at the time was that only a major structural change in the Finnish agricultural sector could bring a viable and long-term solution to the problem. It was estimated that only half of the agricultural holdings were economically viable, and could provide an adequate standard of living to the farming population. Consequently, the merger of small farms into bigger and more productive production units has been encouraged. As a result, the number of agricultural units above 50 hectares has increased substantially; they represented 8.8 per cent of all units in 1997, against 6.8 per cent in 1995 (CEC, 2001).

Since the three new applicant members had a GDP per capita substantially higher than the corresponding average EU figure, and given the structure of their agricultural sector, the impact of the fourth enlargement in budgetary terms meant that these countries were net contributors to the EC budget. For example, it was calculated that Finland's accession would increase community budget expenditure by ECU 830 million, whereas the total income generated by its accession would be higher, at ECU 945 million. In the same vein, the revenue induced by both Austria's and Sweden's accessions was greater than their respective 'costs' to the budget. An extremely different and contrasting picture emerges with the enlargement to the CEECs.

6.2 Enlargement to the CEECs, to Cyprus and Malta

It can be argued that the reunification of the two Germanys and the incorporation of East Germany into the CAP in 1991 provided for the first experience of integration with the former communist world.

However, although the inclusion of East Germany into the EU had also notable implications in terms of the philosophy of EU financing, the magnitude, the nature as well as the importance of the enlargement to central and Eastern Europe is totally unprecedented. This section is organized as follows: we will first present the applicants' path to accession; second, we will propose a short synopsis of the agricultural sectors of the applicant member countries; finally, we will discuss the implications of the enlargement for both sides.

6.2.1 The path to accession

From the early 1990s, most CEECs and Baltic states expressed a desire to join the EU. As of late 2002, 13 countries had applied to become members of the EU, namely: Bulgaria, the Czech Republic, Estonia, Hungary, Latvia, Lithuania, Poland, Romania, Slovakia, Slovenia, Cyprus, Malta and Turkey. This is undeniably a very heterogeneous group of countries, combining small and large states, formerly centrally planned and market-led economies, Mediterranean and North European states. Given the size of both Malta and Cyprus, as well as their increasing closer ties with the EU through preferential agreements, the accession of these two countries will have no more than minor implications.[2] Turkey, which officially applied in 1987, is the least likely country to join in the near future (see below). This is why the analysis in this chapter will focus mostly on the experience of the ten CEECs.

Bilateral negotiations, between the EU on the one side and each applicant country on the other, with the exception of Turkey given that it does not meet the political conditions as laid down in the so-called 'Copenhagen criteria', started in the late 1990s. The first wave of enlargement will see a large number of applicant countries join the EU as early as 2004; these ten forerunners are: the Czech Republic, Estonia, Hungary, Poland, Slovenia, Slovakia, Latvia, Lithuania, Cyprus and Malta. The date of accession for Romania and Bulgaria depends on the progress made on a number of socio-political and economic issues and on the pace of negotiations. The bilateral negotiations cover 31 chapters in total, that is all areas of the *acquis communautaire*. The *acquis* refers to every possible ingredient of European integration achieved so far by the

EU countries. The 31 chapters deal therefore with the different constituent elements of European integration, such as science and research policy, EMU (Economic and Monetary Union), the free movement of capital, transport, energy, and so on. Agriculture features therefore as one of the important chapters (Chapter 7). Negotiations on Chapter 7 started in June 2000 with the so-called 'Luxembourg group', comprising Hungary, Poland, the Czech Republic, Estonia, Slovenia and Cyprus. Further negotiations with the 'Helsinki Six', that is with Slovakia, Romania, Bulgaria, Latvia, Malta and Lithuania opened gradually in June 2001. However, eager to prepare their *rapprochement* with the EU from the onset, the CEECs had already ventured into a number of negotiations in the early 1990s. These negotiations culminated with the 'Europe Agreements'; the 'Europe Agreements' can be seen as a step towards free trade with the EU, and therefore as a preparation towards full membership.

6.2.1 Europe agreements

A first round of Europe agreements was signed in December 1991 with Hungary, Poland and Czechoslovakia. The agreement with Czechoslovakia was not ratified, since the country split into two independent states. Consequently, the agreement was then re-negotiated with the two independent entities, and two separate agreements were signed in 1993. Similar agreements were signed in 1993 with Bulgaria and Romania, in 1995 with the Baltic States and in 1996 with Slovenia. Behind the emphasis on the need for political dialogue, the Europe agreements aimed at liberalizing trade between the two Europes. However, although the agreements covered economic co-operation and trade, they did not include all the agricultural and fisheries products, judged 'sensitive' by the EU commission. In particular, high trade barriers were kept in fisheries, a vital industry for Poland, and in wine, an important sector for Hungarian agriculture. As documented by Maresceau (1997), the Europe agreements were not originally seen, by the Commission, as pre-accession agreements. It is only in 1993 that a political reorientation of these agreements transformed them into a basis for a pre-accession strategy, or transition process. As defined by the 1994 Essen European Council, the three elements of the pre-accession strategy, which offers support to all applicants irrespective of their current state of preparation, are: the Europe Agreements, a Structured Dialogue and PHARE. The PHARE program is an essential element in this strategy, as it represents the main source of financial assistance before accession. It has been providing financial assistance to the countries of central Europe since 1989, for institution building and for investment projects. Institution building can take the form of technical assistance, training programs as well as exchange of experts.

6.2.1.2 Assessing the suitability of applicant countries – the Copenhagen criteria

However, before the negotiations for accession could start, the different applications for membership were thoroughly examined by the Commission, with reference to a number of criteria. These criteria were adopted by the European Council when it met in Copenhagen in June 1993, and they are today therefore referred to as the 'Copenhagen criteria'. Membership of the EU requires first of all that the candidate country (CEC, 1993):

- has achieved stability of institutions guaranteeing democracy, the rule of law, human rights and respect for and protection of minorities,
- has put in place a market economy and can cope with competitive pressures and market forces within the Union,
- has the ability to take on the obligations of membership, including the adherence to the aims of political, economic and monetary union.

In particular, the ability to cope with competitive pressures requires a sufficient amount of human and physical capital, and the restructuring of state-owned enterprises, including in the agricultural sector. The EU Commission reviews periodically the progress made by each candidate country in meeting the Copenhagen criteria and it prepares opinions on each application for membership at regular intervals. It also makes an analysis of expected progress over the medium term, as well as of the ability of the applicant country to implement the *acquis*. The first such opinion was published in 1997, and since then the Commission has published regular annual reports on each applicant country. For example, in its 2000 Report (CEC, 2000), the Commission found that Turkey had achieved very little progress on the Human Rights issue (criterion number 1), undermining the chances of entering into a negotiating process with this country. At the June 1997 Amsterdam summit, the EU leaders started considering the applications of some of the CEECs, so as to initiate the negotiation process, as described above. Negotiations have subsequently been opened with every candidate country that has met the political criteria for accession and that has shown that adequate measures to meet the economic criteria have been taken.

6.2.1.3 The acquis in agriculture

Implementing the *acquis communautaire* in the area of agriculture implies that the various applicant member states must start reforming their agriculture and food production systems prior to membership, so as to ease subsequently their integration into the EU. For example, Slovenia embarked upon a reform program in 1998, in line with the major elements contained in the CAP. In particular, Slovenia's agricultural reform has four pillars: market price policy, direct payments, a program of restructuring agriculture and food production, and rural development. Since 1998, the Slovenian government has been implementing several national programs related to agriculture; one such program is the 'National Programme for the Adoption of the Acquis'. These programs are aimed at allowing the indigenous agricultural policy come closer to the CAP during the period leading up to accession. More specifically, the strategy consists in adopting the market regulations existing under CAP, in particular those relating to milk and dairy products, beef, pork, fruits and vegetables, grain for bread making, sugar, honey as well as sheep and goats meat, all important sectors for Slovenian agriculture (Slovenian Ministry of Agriculture).

6.2.2 A sketch of the applicants' agricultural sectors

During the Communist era, the agricultural sectors of the CEECs were generally characterized by central planning, large state-controlled agricultural holdings, low levels of technology, low productivity levels, a highly vertically integrated agri-food chain from input providers to final consumers and inadequate food supplies on markets. The agricultural knowledge systems (AKSs) of these countries were often subjected to the political needs of the centrally planned system.

Although these countries share a number of similarities, there are however many significant differences between them. For example, Czechoslovakia, Hungary and Slovenia were countries with higher levels of mechanization. The economic and social heterogeneity of the applicant countries can be gauged from Table 6.1. In the poorest applicant countries, namely Bulgaria, Romania, Lithuania and Latvia, average expenditure on food was still well above 33 per cent of total expenditure in 2000. This compares with an average of approximately 10 per cent in the EU in 1999 (EUROSTAT, 2000). In the case of relatively richer countries such as Slovenia or Malta, the share of total expenditure devoted to food consumption is still 50 per cent above that of high food spending EU countries such as Spain or Italy. There is

Table 6.1 Macroeconomic indicators of the CEECs (2000 or 2001)

	Population (000s) (2001)	GDP/Capita in PPP (% of EU average) (2000)	Expenditure on food* (as a % of total expenditure) (2000)	Share of employment in agriculture (in % of total) (2000)
Bulgaria	8,149	24.1	42.2	13.2
Cyprus	759	82.6	–	5.4
Czech Republic	10,266	60.1	21.3	5.2
Estonia	1,367	38.0	32.7	7.0
Hungary	10,005	52.8	34.9	6.5
Latvia	2,366	29.2	33.3	14.4
Lithuania	3,693	29.5	44.4	18.4
Malta	391	53.2	20.4	1.7
Poland	38,644	38.9	30.8	18.7
Romania	22,430	26.9	38.5	45.2
Slovakia	5,402	48.1	26.2	6.9
Slovenia	1,990	71.6	20.1	9.6
Turkey	65,784	28.6	–	34.9

NB: GDP figures are in current prices. * Food and non-alcoholic beverages.
Source: EUROSTAT (2002) Statistical Yearbook on Candidate and South-East European Countries (European Commission, Brussels and Luxemburg), CEC (2002a).

therefore an appreciably wide gap between food expenditure as a percentage of total expenditure in the applicant countries when compared with the EU countries. For some of the CEECs, decreasing disposable incomes during most of the 1990s meant that the average share of household income spent on food actually increased during the period. This is the case for countries such as Bulgaria, Slovakia and for the three Baltic States (Trzeciak-Duval, 1999).

Another major distinguishing feature between relatively poor and relatively rich countries in the group of applicant economies is given by agricultural employment as a share of total civilian employment. As can be seen from Table 6.1, the decision to absorb large and poorer countries, such as Romania only at a later stage, is understandable for economic reasons. In the year 2000, Cyprus, the Czech Republic, Malta, Hungary and Slovakia, all had smaller agricultural sectors, in relative terms, than Greece, Portugal and Ireland. However, even in a country such as Hungary with a relatively small agricultural sector, in terms of numbers employed, the agri-food complex is an important element of the economy. This country is endowed with a favorable climate, fertile soil, enabling it to produce a diversified range of agricultural products, including cereals, vegetables, fruits, and dairy products, and to raising horses, sheep, poultry, as well as fur-bearing animals. In the second half of the 1990s, the agricultural, food and forestry sectors together accounted for more than 10 per cent of Hungary's GDP, and the agri-food complex represented an estimated 15–18 per cent share of the country's GDP. In the rural areas, agriculture is still the main source of employment in Hungary (Hungarian Ministry of Agriculture, 2000). Generally, in most applicant countries, farming still constitutes an important social safety net in rural areas. It should also be pointed out that because of rising unemployment levels in these countries during the first half of the 1990s, agriculture has tended to absorb surplus labor from the manufacturing sector in some of these countries. Indeed, in Bulgaria, Latvia, Lithuania and Romania agricultural employment in relative terms has in fact increased between 1989 and 1997 (Trzeciak-Duval, 1999).

The 1990s have nevertheless been witness to a great deal of economic change in the CEECs. Agricultural employment, as well as spending on food products as a proportion of total expenditure, is lower today than it was in the early 1990s. The transition to a market-oriented economic system has been greatly helped by the three elements of the pre-accession strategy. During the 1990s, the CEECs have introduced agricultural policies strongly influenced by the CAP mechanisms. Many of these countries now operate a system of agricultural support including price guarantees with intervention buying, export refunds and tariffs (Hertel *et al.*, 1997). For example, agricultural support in Hungary takes the form of market price support subsidies, credit subsidies and export subsidies (CEC, 2002a). Because of the unprecedented changes experienced by the applicant economies from the East since the early 1990s, most of these countries can now be considered as functioning market economies, and, in the eyes of the EC Commission, they should be able to meet the second criterion in the short term.[3]

Privatization programs, foreign direct investment enticing policies and land reforms aimed at re-establishing private ownership in the rural areas, have all permitted the restructuring of agricultural holdings in the CEECs (see for example Csaki and Lerman, 1997). Large cooperatives, or partnership structures of several hundred hectares, were often the legacy of the centrally planned system in many of these countries. In the case of the Czech Republic, only 3 per cent of the total agricultural land was owned and farmed by individual owners in 1989 (CEC, 1996). This share increased to 70 per cent in 2000 (Table 6.2). The cooperatives or partnership structures had an average size of 1200 ha in the case of

Table 6.2 Land by legal status in the CEECs (in %, 1996 and 2000)

	State enterprises		**Cooperatives**		**Others**	
	1996	**2000**	**1996**	**2000**	**1996**	**2000**
Bulgaria	21.0	18.0	–	–	79.0	82.0
Czech Republic	2.1	0.9	37.0	29.1	60.9	70.0
Estonia	1.0	0	27.3	23.1	71.7	76.9
Hungary	17.6	14.5	28.3	15.3	54.1	70.2
Latvia	0.8	0.3	4.8	–	94.4	99.7
Lithuania	0.7	0.5	15.6	3.2	83.7	96.3
Poland	6.7	5.7	2.7	2.0	90.6	92.3
Romania	28.0	11.1	11.0	8.7	61.0	80.2
Slovakia	14.4	1.9	59.0	49.7	26.6	48.4
Slovenia	–	–	14.9	5.8	85.1	94.2

NB: The 'Others' category refers to private farms or individual holdings.
Source: EUROSTAT (2002) Statistical Yearbook on Candidate and South-East European Countries (European Commission, Brussels and Luxemburg).

this country. The situation was however different in other CEECs. In Poland for instance, more than 72 per cent of the land was in the hands of private farmers in 1989; this proportion was above 90 per cent in 1994 (CEC, 1996).

Although a direct contrast between these countries is hampered by the fact that the definitions of state-owned enterprises (SOEs) and cooperatives differ between countries, a comparative assessment of their progress towards land restitution and privatization has been attempted in Table 6.2. SOEs generally refer to enterprises owned and managed by the State, whereas cooperatives are normally funded by several partners/managers who share the profits. In most countries, private ownership of utilized agricultural land is the normal rule by 2000. Slovakia is however a country where cooperatives are still a dominant form of land ownership and management. In other countries such as Bulgaria, land restitution and privatization were completed by the end of 2000 (CEC, 2002a).

The existence of a positive relationship between farm structure and efficiency seems to have been gaining credence since the fall of the Berlin wall. During the first half of the 1990s, Czech family farms were found to be more efficient than both the cooperatives and the commercial companies that succeeded state farms, in the sectors of crop production and livestock fattening. In the late 1990s, total factor productivity (TFP) for family farms, defined as $Q/(\alpha K + \beta L)$ was considerably higher than the TFP for cooperatives (Sarris *et al.*, 1999).

Extensive restructuring during the post-Communist years has nevertheless coincided with important decreases of agricultural output in volume terms in all CEECs. The annual growth in volume of agricultural output has been negative well into the second half of the 1990s, a trend that contrasts with manufacturing output over the same period. For example, the Hungarian agricultural sector has suffered substantial setbacks during the 1990s. These are clearly visible in the livestock sector where the number of cattle and pigs has declined by nearly 10 per cent between 1996 and 2000 (EUROSTAT, 2002). Hungary's current levels of agricultural and food production are well below the sector's potential output, and the government is actively trying to stimulate the industry's recovery, by placing an important emphasis on agriculture and rural development. These sharp declines in agricultural output are explained by a number of factors, among which: the loss of traditional CMEA markets, matched by a slow trade reorientation process; the fall in disposable incomes; uncertainty in the agricultural sector brought about by the transition to a market economy; and the shortage of technical and management skills (Trzeciak-Duval, 1999). A small recovery of gross agricultural production in volume terms has nevertheless been registered at the close of the millennium. In countries such as Latvia, Lithuania and Slovenia, gross agricultural output has increased in 2000 compared with the previous year (EUROSTAT, 2002). It is however too early to infer any trends from this small positive performance.

6.2.3 Benefits and costs arising from the fifth enlargement

6.2.3.1 Benefits

One of the major benefits enjoyed in the post-Communist years by the CEECs, has been their ability to re-direct their trade away from the old CMEA (Council for Mutual Economic Assistance) structure and towards Western Europe. Although Russia is still today a relatively important trade partner for countries such as Lithuania, Bulgaria and Slovakia, the EU as a whole represents nevertheless the major source for and the major destination of trade in general and of agricultural trade for these countries. For the applicant economies, trading with economically advanced countries represents various advantages such as the ability to import a larger spectrum of diversified products. In particular, trade reorientation has been beneficial to a country such as Hungary. In 1997, nearly a quarter of Hungarian food production was exported to countries such as Germany, the largest absorber of Hungarian food products, as well as to other neighboring countries such as Russia, Italy,

Poland and Austria (Hungarian Ministry of Agriculture, 2000). Among the other benefits arising from the fifth enlargement are the prospects for rapid economic growth in the medium to long term in the CEECs. Rapid growth combined with rising incomes should translate into higher demand for EU differentiated products. The EU agricultural sector has already made successful inroads into the CEECs markets, both through exports and through intensified foreign direct investment in food and agriculture. Between 1995 and 1998, the EU's agricultural exports to the CEEC-10 increased by 25 per cent, whereas its imports from the East European countries increased by only 13 per cent. This has contributed to enlarge the trade surplus that the EU enjoys with most of the CEECs. Table 6.3 shows that Hungary is the only exception to this rule in that it had a trade surplus with the EU of more than ECU 500 million in 1998. In spite of lower food export levels in the 1990s, Hungary has been able to sustain its trade surplus with the EU over the period. Every other East European country is dependent on the EU for the supply of food products. The dependence is particularly acute for some of the Baltic States such as Latvia, Estonia, but also for Slovenia. The nature of the main agricultural products traded between the two

Table 6.3 Bilateral agricultural trade between the EU and each of the CEECs (1998)

Country	**Trade balance (Mio ECU)**	**Major EU products exported**	**Major products imported by the EU**
Poland	721	Residue and waste from food industry	Edible fruits and nuts, live animals
Czech Republic	653	Residue and waste from food industry	Beverages, spirits and vinegar
Hungary	−520	Residue and waste from food industry	Meat
Slovenia	325	Tobacco	Meat
Estonia	238	Beverages, spirits, vinegar	Dairy products, eggs, natural honey
Slovakia	195	Residue and waste from food industry	Residue and waste from food industry
Romania	235	Tobacco	Live animals
Bulgaria	9	Meat	Beverages, spirits and vinegar
Latvia	187	Animal or vegetable fats and oils	Dairy products, eggs, natural honey
Lithuania	207	Residue and waste from food industry	Dairy products, eggs, natural honey

Source: CEC (2001).

Europes reveals a trade dichotomy between the two Europes (Table 6.3). The EU uses some of the CEECs as a source of supply in meat and in other relatively income elastic products such as fruits and nuts, dairy products, eggs and natural honey. This is in contrast to the residues and waste from the food industry, which are the main EU products imported by Poland, the Czech Republic, Hungary, Slovakia and Lithuania.

As was mentioned in Chapter 5, trade and FDI are both essential vehicles of knowledge diffusion in host countries. In particular FDI contributes to structural change in recipient countries by boosting the productivity and efficiency of their industries. Case studies on FDI in the agri-food sector in the CEECs show that the introduction of innovations at the managerial and productive levels has generated high efficiency gains during the 1990s (Gow and Swinnen, 1998). The acquisition of foreign knowledge helps the various AKSs to adapt gradually, so as to engender sustainable increases in agricultural productivity. In recent years, the AKSs of these countries have been substantially reformed; Universities are now more involved in agricultural research and these countries have developed many research linkages with the EU and with other parts of the world (Csaki, 1998).

6.2.3.2 The costs of enlargement

Table 6.4 presents comparative data relating to the various enlargements so far, thus placing the fifth enlargement into a broader historical perspective. Enlargement to the CEECs will add more than 100 million consumers, whose average purchasing power will nevertheless be only a third of that of the current consumers in the Union-15 (CEC, 1997). An important feature will be the reduction in the average per capita GDP for the Community as a whole, as can be seen in the

Table 6.4 Impact of successive enlargements of the EU (based on 1995 data)

	Increase in area %	**Increase in population %**	**Increase in total GDP %**	**Change in per capita GDP (%)**	**Average per capita GDP (EUR-6 = 100)**
EUR-9/EUR-6	31	32	29	−3	97
EUR-12/EUR-9	48	22	15	−6	91
EU-15/EUR-12	43	11	8	−3	89
EU-26/EU-15	34	29	9	−16	75

NB: EU-15/EUR-12 includes Germany reunified.
NB: EU-26 comprises the 10 CEECs and Cyprus.
Source: CEC (1997).

table. This is unprecedented and this is greater than that resulting from the previous enlargements put together. Given their limited contributive capacity to the EC budget, the new applicants will be net beneficiaries. In particular, they will avail of substantial transfers under the various structural policies. For example, enlargement leads to a sharp increase in the population eligible for assistance under Objective 1, which will rise from 94 million to 200 million, representing roughly 60 per cent of the total EU enlarged population. Community research and technology programs are gradually opening to the new applicants within the framework of the pre-accession strategy. Acceding countries will also be eligible for loans by the EIB (European Investment Bank) and the EURATOM, on equal terms with existing member states.

With regard to agriculture proper, the EU agricultural area will increase by half and the agricultural labor force would at least double. From the EU's point of view, the most important impact of the new enlargement is in terms of the implications on the EC budget. Although increasing expenditure will be felt at all levels, for example enlargement will add to the Institutions' administrative costs, the major impact will be visible at the level of the CAP and of the structural policies.

6.2.3.3 Impact on the EC budget

Given the limited contributive capacity of the new members to the EC budget, and given the size and structure of their agricultural sectors, the fifth enlargement would, *ceteris paribus,* have enormous budgetary implications. However, as we have seen in Chapter 4, the 1992 Edinburgh European Council decided on a tight budgetary discipline for the decade. This implies that spending on agriculture be curbed by limiting the annual rate of increase in the expenditure concerned. Support prices in the EU have progressively decreased so as to eventually come in line with world market prices (see Chapter 8). Given the reluctance of the major contributors to the EC budget to increase substantially the revenue of the budget, it is clear that measures such as price decreases, are the only possible means to lessen the shock of enlargement to the EU finances.

The principle of a budgetary discipline has helped produce an early estimate of the budgetary implications of enlargement towards Eastern Europe. This was contained in a document that was presented by the then President of the Commission Jacques Santer in June 1997 to the European Parliament. This document, known as 'Agenda 2000' includes the estimated costs to be incurred by the enlargement into a financial perspective covering the time frame 2000–06. This document, agreed by the Council at the 1999 Summit in Berlin, must be seen as an important step in the CAP reform process. All estimates are based on the assumption that a first wave of accessions will take place sometime shortly after 2002. These estimates have been used as policy guidelines for a further strict control of budgetary, and in particular agricultural, expenditure in the EU.

As can be seen from Table 6.5, agricultural expenditure in 2006 would have amounted to €50 billion, had the EU decided not to incorporate the new members, or to delay the enlargement. However, with the new Central and East European applicant members, agricultural expenditure will increase to €54.5 billion, that is a 9 per cent increase. The difference between agricultural spending with and without the new members represents roughly 8.3 per cent of the total agricultural spending in that year. Similar calculations have been made for the structural funds. Further estimates of the budgetary costs of enlargement based on the assumption that up to ten countries would join the EU in 2004 have shown that the extra costs related to enlargement can be accommodated within the Berlin agreement (CEC, 2002b). Costs under the 'CAP' heading include for example direct payments and storage costs (see below). Recent projections show that surplus production could develop in areas such as cereals, oilseeds and pig meat, until 2006 (CEC, 2002b).

The key point in this financial perspective is that the financing of the enlargement can be achieved within an unchanged ceiling of own resources as a percentage of EU GDP (that is 1.27 per cent). This rigorous principle is the guiding element in the financial framework 2000–06 (CEC, 1997). The principle of budgetary discipline has been subsequently

Table 6.5 Expenditure on agriculture and on structural operations (ECU billion)

Agricultural expenditures*	**2002**	**2006**	**Structural funds****	**2002**	**2006**
EU-15	47.4	50.0	EU-15	34.2	30.2
New members			New members		
CAP	1.1	1.4	(incl. cohesion funds)	3.6	11.6
Rural development	0.6	2.5			
Pre-accession aid	0.6	0.6	Pre-accession aid	1.0	1.0
Total agriculture	49.8	54.5	Total	38.8	42.8
New (%)	4.6	8.3	New (%)	11.9	29.4

* Current prices.
** 1997 prices.
Source: CEC (1997).

echoed at the 1999 Berlin Summit. The conclusions of the European Council in Berlin include the Commission's proposed guidelines of EAGGF guarantee expenditure as well as two sub-guideline ceilings: one for traditional market expenditure and the other one for expenditure on rural development.

As was mentioned above, the PHARE program is an essential financial tool in the pre-accession strategy. The CEECs received ECU 6.7 billion under PHARE between 1995 and 1999. In addition to PHARE, a new pre-accession aid of ECU 1.6 billion per year has been granted from the year 2000 onwards to the applicant CEECs, of which ECU 0.6 billion is earmarked for agricultural development (Table 6.5). The Council of Ministers decided to reinforce pre-accession assistance in favor of agriculture and rural development, by creating a specific instrument called SAPARD (Special Accession Programme for Agriculture and Rural Development). Covering again the period 2000–2006, SAPARD was allocated a budget of €520 million per year. The idea of SAPARD is to help the applicant countries implement the *acquis communautaire* under CAP and other related policies (CEC, 2001). Funding under SAPARD is aimed at modernizing agricultural holdings and structures in these countries, and it will cease after the accession of the applicant countries. SAPARD funding can be utilized for the following purposes: investments in agricultural holdings; investments in the processing and marketing of food products; investments to enhance foodstuffs quality and consumer protection; land improvement and re-parceling; water resource management, and so on (CEC, 2001). Table 6.6 gives an approximate idea of the breakdown by country.

Table 6.6 Funding under SAPARD (at 1999 prices)

Country	**Annual amount (€ million)**
Bulgaria	52.1
Czech Republic	22.1
Estonia	12.1
Hungary	38.1
Lithuania	29.8
Latvia	21.8
Poland	168.7
Romania	150.6
Slovenia	6.3
Slovak Republic	18.3

Source: CEC (2001).

Larger amounts of funding are allocated to those countries with greatest needs, and as can be seen in Table 6.6, the two largest countries, Poland and Romania, are allocated more than half of the total funding on an annual basis.

6.2.4 Transposing the *acquis* to the new applicant countries

All negotiating issues covered by Chapter 7 in terms of direct payments, production quotas as well as rural development were addressed by the EU during the first half of 2002. The level of direct payments and production quotas obtained by the CEECs and other applicant economies is crucial for the future of their agricultural sector. Rural development policies are essential for guaranteeing a balanced economic development in the rural areas that tend to be the poorer areas in most countries.

6.2.4.1 Direct payments

As was seen in more detail in Chapter 4, direct payments (or direct aid) were introduced in the aftermath of the support price cuts of the 1992 and Agenda 2000 reforms, in order to compensate for the loss in farmers' income in the EU. Granted to EU farmers in the case of arable crops and cattle, they were introduced in 1995 in the rice sector, and they will extend to milk from 2005 onwards. Direct payments are part of the CAP *acquis*, and they should therefore be extended to applicants so as to abide by the principle of a single market for agricultural products which is the cornerstone of the CAP. Applicant countries have indeed requested the same treatment as EU countries with regard to this matter.

However, it is felt that the effects resulting from a rapid application of EU prices to candidate countries would be manifold and counterproductive. For example, the introduction of direct payments in line with those paid to EU farmers could encourage cereal production, or/and may lead to a faster development of specialized beef production. If direct payments are too high and if they are introduced too quickly, there is a risk that 'necessary restructuring will be slowed down or even stopped, creating a durable vicious circle of low productivity, low standards and high hidden unemployment' (CEC, 2002b: 5). Also, by favoring certain groups of farmers, they could create income disparities and substantial social distortions in rural areas, a problem that developed at EU level. Therefore, the level of and timing for the introduction of these payments are crucial points in the negotiations. As a result, the preferred option for the Commission is to have them introduced both at a low level and gradually during a transition period of ten years, period after which the direct aids would have reached the EU level.

During the ten-year transition period, the Commission advocates a compensation through intensified financial support for restructuring (CEC, 2002b). The merit of this direct payment approach is that it can be realized under the currently applicable WTO arrangements.

However, and as stated above, most applicant countries have prepared intensively for accession. One feature of this preparation is visible through the producers' food prices. These prices have increased in recent years in applicant countries so as to come into line with EU prices. Although part of this increase can be explained by the positive growth rates and economic recovery experienced by these countries since the second half of the 1990s, some of this increase can nevertheless be attributed to the introduction, in some applicant countries, of 'CAP-like' direct payments to the benefit of farmers.[4] There has been indeed a gradual shift away from price support to direct payments in the CEECs. It should however be noted that after accession, these direct payments will in fact be the equivalent of national aids, and that they may be incompatible with the *acquis* on state aids in agriculture.

6.2.4.2 Production quotas

Production quotas in the milk, sugar and arable crops sectors are calculated on the basis of historical production data. The determination of quotas for the applicant countries is made difficult by the arbitrariness, which exists when deciding the period of reference to use for calculation purposes. The Commission's view is that the 1995–2000 period should be used as the reference period for the calculation of quotas (CEC, 2002b). The justification for this choice lies in the fact that the economic conditions experienced by the CEECs in the 1980s were atypical and very different from those experienced subsequently. Moreover, the early 1990s were unusual, for they were a period marked by many problems arising from the transition period. Other reasons advocated by the Commission include the more reliable statistical apparatus of CEECs in recent years.

The negotiations are generally made complex and sensitive by the often significant gap existing between the requests from applicant members and the recommendations of the Commission for each of the measures. With regard to the determination of quotas, there are ample divergences between applicant member states and Commission. For example, the milk and iso-glucose quotas requested by the various applicant countries and those proposed or considered by the Commission vary substantially. In many instances, the requests expressed by the candidate countries are found by the Commission to be higher than is justified by the historical data.

6.2.4.3 Rural development

Restructuring *per se* is not sufficient to address the problems of the CEECs' rural communities in the future. It could generate growing rural unemployment and poverty without being able to provide alternative sources of income directly. The Commission acknowledges the need for a safety net, an issue that has started being addressed through the pre-accession instrument SAPARD. Moreover, as most applicant countries will be classified as Objective 1 regions, much of their rural policy measures will attract funding from the EAGGF guidance and from other sections of the structural funds.

6.2.5 Other impacts

6.2.5.1 Producer and consumer prices

Price liberalization in the early 1990s led to unprecedented inflation levels in the CEECs. In the first half of the 1990s, many countries experienced three-digit inflation rates. Between 1989 and 1992, prices for farm inputs reached world market levels. As no compensatory changes in producer prices were introduced to offset these increases, farmers suffered harsh price squeezes (Trzeciak-Duval, 1999). In subsequent years, price stability was achieved in most countries, with the exception of Bulgaria and Romania. Because of the new system of agricultural support introduced in these economies, with price guarantees leaving way progressively to direct compensation, producer prices are normally above world market prices, although they still remain below EU prices. However, in recent years, they have tended to increase so as to converge towards (decreasing) EU price levels. Since food expenditure still represents an important share of household expenditure in many of the CEECs, an abrupt increase in consumer prices to EU price levels could exacerbate social problems and undermine the efforts made to maintain a certain level of social stability and cohesion in these countries.

6.2.5.2 Impact on other EU countries

Modern modeling techniques enable the quantification of the possible trade, investment, and other macroeconomic impacts arising from the fifth enlargement on each EU country, and many studies have endeavored to quantify these effects. For each country, the study of this impact can be effectuated using partial or general equilibrium models. Partial equilibrium analysis involves measuring the effects of enlargement on a specific industry or sector, for example the agricultural sector, whereas general equilibrium models quantify the effects on all industries. Gravity models tend to suggest that those countries that are geographically in the vicinity of the CEEC would be most affected by the enlargement. Austria is one such country. In a recent study, Kohler and Keuschnigg (1999) evaluate the potential costs and benefits of Eastern enlargement to the EU in general and to Austria in particular. Their conclusions are that the budgetary costs of enlargement are significantly outweighed by the dynamic gains arising from integration, and that, in short 'extending EU membership to Eastern applicants is more than worth its price for Austria' (1999: 1), although there may be losses in the agricultural sector *per se*. The dynamic gains from integration encompass gains in terms of structural change, economies of scale and economic efficiency. What a study such as this one suggests is that the impact of the enlargement should be best appraised at the level of the economic system of a country as a whole. Projections by the Commission show that in the milk and beef sectors many of the applicant countries would become net importers during this decade, as consumer disposable income and demand grow (CEC, 2002b). The potential beneficiaries of these developments are EU countries such as Ireland, for beef, and The Netherlands for dairy products.

Summary

The fifth enlargement expands the EU to include another ten European countries, that is Cyprus, the Czech Republic, Estonia, Hungary, Latvia, Lithuania, Malta, Poland, Slovakia and Slovenia. The entry of Bulgaria and of Romania in the EU is delayed, on political grounds, until the end of the decade. Enlargement to all the CEECs, to Malta and Cyprus would increase the total agricultural area in the EU by half and would multiply the agricultural labor force by two. The new applicant countries have levels of GDP per capita substantially lower than the EU average; their agri-food *filière* is characterized by low levels of capitalization, innovation and therefore of productivity, calling for major structural transformations. Structural change has been initiated from the early 1990s, when the CEECs signed a number of Europe agreements with the EU. Trade reorientation, in favor of the EU, has given an impetus to the reform of their economic and agricultural systems. These reforms enable these countries to embrace gradually the *acquis*. The major impact of the enlargement on the EU system is felt at the level of its finances.

Agenda 2000 estimates the costs incurred by the enlargement over the time frame 2000–06. The principle of budgetary discipline allows for the enlargement to take place within an unchanged ceiling of own resources as a percentage of EU GDP. For the new member countries, benefits of the enlargement include economic growth, through the absorption of new imported technology, the ability to import EU differentiated products, and the modernization of farms and food industries. However, one of the major challenges for these countries is to contain the food price rises within reasonable limits, as well as to guarantee a balanced economic development in rural areas.

Key terms and concepts

Centrally planned economic system
Structural change
Europe agreements
Acquis communautaire
Pre-accession strategy
PHARE
Copenhagen criteria
SAPARD
Costs and benefits of the enlargement

Notes

1 The European Free Trade Area (EFTA) was created in 1960 by Austria, Iceland, Liechtenstein, Norway, Sweden, Switzerland, Portugal and the UK. A plan to form an Economic Community among the Nordic countries, NORDEC, did not materialize in the 1950s because the Soviet Union resisted Finland's membership. Finland joined EFTA in 1961 as an associated member, and it became a full member of EFTA in 1986. Created by the Luxembourg declaration of 1984, the new concept of European Economic Area, encompassing the EC and EFTA, became reality in January 1994, when the new EEA agreement entered into force. This agreement created the largest and most integrated economic area in the world.
2 Note that Malta had been considered as a candidate for accession in July 1990 and prepared for accession until 1995, when the new labor led Government withdrew its candidacy.
3 The exceptions to this rule are Bulgaria and Romania, the latter country still having to solve human rights problems. Consequently, accession for these two countries will be delayed until January 2007.
4 Note that in the mid-1990s, most CEECs, with the exception of Poland and Slovenia had already reached top limits on domestic support, measured on the basis of Aggregate Measure of Support (Knaster-Sánchez, 1996).

Bibliography

CEC (1993) *Bulletin of the European Communities,* Vol. 26, No. 6, 12–13.

CEC (1996) *The CAP and Enlargement – Economic Effects of the Compensatory Payments.* European Economy, No. 2, Brussels.

CEC (1997) *AGENDA 2000 – For a stronger and wider Union* (Brussels and Luxemburg).

CEC (2000) *Report on Progress Towards Accession by each of the Candidate Countries,* COM (2000) 700 final, Brussels.

CEC (2001) *The Agricultural Situation in the European Union,* 1999 Report (Brussels and Luxembourg).

CEC (2002a) *Evaluation of the 2001 pre-accession programmes of candidate countries,* European Economy, Enlargement Papers, No. 7, January, Brussels.

CEC (2002b) *Enlargement and Agriculture: Successfully integrating the new Member States into the CAP,* SEC (2002) 95 final, Brussels.

Csaki, C. and Lerman, Z. (1997) 'Land reform and farm restructuring in East Central Europe and CIS in the 1990's: Expectations and achievements after the first five years', *European Review of Agricultural Economics,* 24: 3–4, 428–52.

Csaki Csaba (1998) 'Agricultural research in transforming Central and Eastern Europe', *European Review of Agricultural Economics,* 25: 3, 289–306.

EUROSTAT (2000) *EUROSTAT Yearbook 2002 – The Statistical Guide to Europe; Data 1990–2000,* Luxembourg.

EUROSTAT (2002) *Statistical Yearbook on Candidate and South-East European Countries* (Brussels and Luxembourg, European Commission).

Gow, H.R. and Swinnen, J.F.M. (1998) 'Up- and downstream restructuring, foreign direct investment and hold-up problems in agricultural transition', *European Review of Agricultural Economics,* 25: 3, 331–50.

Hertel, T.W., Martina, B. and Padma V. Swaminathan (1997) 'Sectoral and economy-wide analysis of integrating Central and Eastern European countries into the EU: Implications of alternative strategies', *European Review of Agricultural Economics,* 24: 3/4, 359–89.

Hungarian Ministry of Agriculture (2000), www.gak.hu.

Kohler, W. and Keuschnigg, C. (1999) *Eastern Enlargement of the EU: How Much is it Worth for Austria?,* Working Paper Series, University of Linz and University of St Gallen.

Knaster-Sánchez, B. (1996) *Challenges for Europe's Agricultural Policy* (Centre for European Policy Studies Review, No. 1, Summer, 8–12, Brussels).

Maresceau Marc (ed.) (1997) *Enlarging the European Union – Relations between the EU and Central and Eastern Europe* (London: Longman).

Sarris, A.H., Tomáš, D. and Erik, M. (1999) 'Agricultural restructuring in Central and Eastern Europe: implications for competitiveness and rural development', *European Review of Agricultural Economics,* 26: 3, 305–29.

Trzeciak-Duval, A. (1999) 'A decade of transition in Central and Eastern European agriculture', *European Review of Agricultural Economics,* 26: 3, 283–304.

7
The EU as an Agricultural Trade Partner

Objectives of this chapter

- To provide an overview of world trade patterns in agriculture, and to discuss the comparative advantage theory.
- To analyze the role of the EU in world agricultural trade.
- To study the EU trade agreements with a number of selected developing countries, such as: ACP (Africa, Caribbean and Pacific) countries, MEDA (Mediterranean Countries), and ASEAN (Association of South East Asian Nations) and other Asian countries.
- To review briefly the issue of rural poverty and malnutrition in the developing world, and to examine the role the EU can play through its co-operation and food aid programs with the least developed countries (LDCs).

Introduction: overview of world trade patterns and comparative advantage of countries

Increasing trade across nations of the world has been one of the dominant features of post-war economic development. In the last 40 years, the share of world trade in total world GDP has more than doubled, increasing from a quarter of GDP in 1960 to nearly 52 per cent in 1999 (World Bank, 2001). In the agricultural sector, food exports have also increased faster than food output. This increase in volume as well as in value terms has greatly been facilitated by the establishment of the General Agreement on Tariffs and Trade (GATT) in the late 1940s, whose aims have been continued by the WTO (World Trade Organization). The way in which the GATT/WTO has enabled a greater economic integration of countries through trade will be examined in Chapter 8.

In spite of their inexorable declining trend, agricultural products represented 7.6 per cent of total world exports in 1998; the corresponding figure for the EU-15 was 7.0 per cent (CEC, 2002). However, for many countries of the world, exports of food products are still an important means of increasing both incomes and foreign exchange earnings. This may lead to the well-known problem of heavy reliance on agriculture, and of a non-sufficiently diversified export structure. In Europe for example, Cyprus provides the example of a country heavily dependent on agricultural trade. In 2000, food and drinks accounted for more than a third of its exports in commodities, although this share has tended to decline over the 1990s (EUROSTAT, 2002). Anderson (1988) notes that, for some countries, a marked comparative advantage in exporting agricultural products has slowed down the inexorable decline of the agricultural sector, as was described in Chapter 1.

The principal agricultural products in world production and trade

As can be seen from Table 7.1, cereals (with the exception of rice), feed grain, oilseeds, milk, meat and sugar are the major commodities in terms of world production. The table shows the important production shift that has been occurring during the 1990s, with a substantial increase in the production of meat products, feed grain and oilseeds worldwide. The last column of the table suggests that the reordering of

Table 7.1 World production and trade in the principal agricultural commodities (1991 and 1998)

	World production (1000t)		**World trade as a % of world production (in volume terms)**		**Net EU share of world trade***	
	1991	**1998**	**1991**	**1998**	**1991**	**1998**
Total cereals#	1,406,959	1,491,180	13.9	12.7	13.3	17.1
Feed grain#	839,397	902,338	11.1	9.9	6.9	4.3
Total milk	468,559	466,347	0.2	0.2	30.9	17.3
Oilseeds	260,689	330,072	13.5	15.0	42.6	−36.9
Meat	179,594	216,201	4.2	7.0	12.2	9.5
Sugar	125,810	137,685	22.8	26.6	11.5	12.6

* Defined as the % share of the EU in world exports minus the % share of the EU in world imports.
# Except rice.
Source: CEC (2002a).

world agricultural production in the 1990s has been beneficial to the rest of the world, rather than to the EU. Indeed, the EU share of world imports has been increasing much faster than its share of world exports for all the commodities shown in the table, with the exception of sugar. In other words, the EU has been losing ground on all major agricultural export markets, except sugar. Its deteriorating position is particularly noticeable in the case of oilseeds.

Although international trade in agricultural products grew rapidly since the 1970s, only about a quarter in volume terms of world food production was actually traded in the late 1990s. The major reasons explaining this *état de fait* are: the perishable nature of some food products and the traditionally high level of protectionism in the agricultural markets of the developed world. Understandably, commodities such as liquid milk are only marginally traded between the various regions of the world, as can be seen from the low trade share in total production in Table 7.1. Agricultural trade liberalization in the 1990s has clearly been beneficial to some agricultural products such as oilseeds, meat and sugar. Consequently, countries specialized in these products are bound to have benefited from intensified trade. Oilseeds and meat are key products for the USA. In particular, oilseeds accounted for more than 10 per cent of the US food exports in the late 1990s (see Table 7.2). Meat and to a lesser extent, sugar, represent two major agricultural products for the EU in terms of both production and trade. The EU's net share of sugar in world trade stood at around 10 per cent in 1998. Although the EU and USA represent only a quarter of world food production, the two regions play nevertheless a primordial role on the international agricultural market. In 1998, they represented 59 per cent of world trade in food products (OECD, 2000).

Table 7.2 Main food exports by the EU-15 and the USA in 1998

EU-15		USA	
Product	**Mio US$**	**Product**	**Mio US$**
Beverages – spirits	11,029	Cereals	10,207
Dairy products	5,005	Tobacco	6,286
Meat	3,683	Oilseeds	6,319
Fats and oils*	3,613	Meat	5,890
Total – agricultural products	57,914	Total – agricultural products	57,436

* Of animal or vegetable origin.
Source: CEC (2002a).

As can be seen from Table 7.2, the EU and the USA have an equivalent size in terms of exports. The striking differentiating feature between the two regions is the fact that the EU has a more diversified export structure than the USA. The four major food products exported by the EU account for 40 per cent of EU total exports, against 50 per cent in the case of the USA. The reasons for a more intense export specialization in the case of the USA are manifold. Indeed, there are a number of natural and man-made reasons that explain the nature of export specialization for any country in the world. The most widely accepted theoretical framework is Ricardo's comparative advantage theory, to which we now turn.

Ricardo's theory of comparative advantage

Ricardo's theory of comparative advantage shows, with simple algebra, that countries gain through trade (Ricardo, 1817). The British moral philosopher argued that countries specialize through trade, and that the nature of trade specialization is based on the comparative advantage a nation holds *vis-à-vis* other competing nations. Going beyond Adam Smith's theory of absolute advantage (Smith, 1776), Ricardo shows that even when a country presents absolute disadvantages in all industries, in terms of production costs, trade is still worthwhile, as it will enable the country to increase its welfare through specialization. The concept of comparative advantage rests on the idea that countries are different, in the sense that they have different natural resources, as well as different technologies, skills and know-how. Inspired by the Methuen Treaty signed in 1703 between Great Britain and Portugal, David Ricardo uses the example of these two European countries to illustrate his theory. Placed in the economic reality of early 19th century Europe, Ricardo's model is based on a number of implicit (and simplistic) assumptions, such as:

(i) The two countries P and E have *different* resource endowments.
(ii) The factors of production (capital and labor) are immobile internationally, but mobile at the level of the country, whereas goods are perfectly mobile internationally.
(iii) The factors of production are fully utilized, and cannot be increased.
(iv) Perfect competition prevails at the national level (that is there is atomicity, fluidity, product homogeneity, rational agents, and free entry into and exit from the market).
(v) The working force, capital stock and technology are given.

Table 7.3(a) Cost comparisons

	Labor cost of production (men per year)	
	1 unit of wine	**1 unit of cloth**
Portugal	80	90
England	120	100

(vi) Transport costs and any barriers to trade (such as tariffs, quotas or NTBs) are ignored.
(vii) There are no dynamic effects such as changing tastes and economies of scale.

The two country–two product model can be presented as follows: suppose that the costs of producing one unit of wine and one unit of cloth in the two countries are represented as in Table 7.3(a).

The different costs of production are explained, according to Ricardo, by the availability of different techniques of production in the two countries. As can be seen in Table 7.3(a), Portugal has an 'absolute' cost advantage in both industries, since its labor costs for the production of both products are lower than in England. Nevertheless, having an absolute cost advantage in both industries is compatible with having a relative or comparable cost disadvantage in one of the two industries. Ricardo argues that England should specialize in the production of cloth, given the fact that its cost disadvantage is smaller for cloth than for wine. Introducing the notion of opportunity cost, we can state that a comparative advantage exists when one country can produce a product at a lower cost in terms of opportunities foregone, than any other country.[1] The opportunity cost of a product X is simply defined as the amount of another product Y, that has to be given up in order to produce an additional unit of X. The opportunity cost of production for both countries, and in both industries, is given in Table 7.3(b). In

Table 7.3(b) Opportunity costs of production

	Opportunity cost of production	
	1 unit of wine	**1 unit of cloth**
Portugal	80/90 = 0.88	90/80 = 1.125
England	120/100 = 1.2	100/120 = 0.83

Portugal, the cost of production of one unit of wine represents 88 per cent (80/90) of the cost of production of one unit of cloth in that country. Therefore, the opportunity cost of producing one unit of wine in Portugal is 0.88 units of cloth, against 1.2 in England.

Since the opportunity cost of producing one unit of wine is lower in Portugal than in England (that is since $0.88 < 1.2$), it follows that by producing an extra unit of wine, England faces a higher opportunity cost than Portugal. The reverse happens in the case of cloth; by producing an extra unit of cloth, England faces a lower opportunity cost than Portugal. Consequently, international specialization will be determined on the basis of the examination of these opportunity costs (or comparative costs). The diagonal of the table, from the top left to the bottom right, shows the 'specialization path' of the two countries. The low ratios obtained mean that Portugal specializes in the production of wine whereas England specializes in the production of cloth. When trading on the international market, any unit of wine produced in Portugal can be traded against 1.2 units of cloth, which is more than 0.88 units, had Portugal decided not to open its borders to foreign products. The same reasoning applies to the case of England. This simple model shows why international trade is preferred to autarky. Specialization, through trade, raises the efficiency of the world economy as a whole, and it produces mutual benefits to the trading nations.

The comparative advantage theory has been criticized on several grounds, in particular on the basis of its very restrictive assumptions. Given the important incidence of imperfect competition, and of intra-industry trade in our economies,[2] the theory is not very helpful in explaining the patterns of modern trade specialization. However, from a historical perspective, a modernized version of Ricardo's theory helps elucidate some of the different specialization patterns in agriculture between say, Northern and Southern European countries, once for example intra-industry trade is factored in as an important element of the amended theory. For example, it goes without saying that a favorable climate in Southern Europe, combined with know-how accumulated over many centuries, explains why EU wine production is mostly concentrated in countries such as France, Italy or Spain.

7.1 The major agricultural trade partners of the EU

Before we look at trade patterns, one should note the declining contribution of EU production in relative terms during the 1990s. The curbs in EU production in the cereals, sugar, beef and dairy sectors have meant that other countries in the world, in particular the developing economies, were able to increase their agricultural output share during the period. The same remark applies to the relative export shares of countries. Those developing countries with more competitive agricultural sectors increased their world market shares over the period; this is the case for example of Brazil and Argentina. The composition of world agricultural trade has changed geographically, implying also greater trade flows between developing countries. Today, some 40 per cent of the developing countries' agricultural exports go to other developing countries. Moreover, the 1990s epitomize an important shift in the structure of world demand for agricultural products. Developing countries have experienced what has been termed 'a livestock revolution', with its well-known consequences on the demand for cereals used as animal feeds (Delgado *et al.,* 1999). In the meantime, developed countries have increased their demand for exotic products, such as tropical fruits.

Table 7.4 below presents a comparison of EU, US and Japanese trade flows in agricultural products (excluding fish), in 2001. The figures relating to the EU do not include intra-EU trade, and all figures are based on custom sources.

Table 7.4 Main agricultural trade partners of the triad (EU, USA and Japan, 2001) (in billion €)

	European Union		**USA**		**Japan**	
	Imports	**Exports**	**Imports**	**Exports**	**Imports**	**Exports**
European Union	–	–	11.74	7.25	4.82	0.13
USA	7.66	10.29	–	–	14.35	0.39
Japan	0.16	3.92	0.41	11.36	–	–
Candidate countries	6.49	6.70	0.77	0.95	0.24	0.00
ASEAN	4.18	2.01	3.59	2.87	3.48	0.21
Other Asia	3.30	2.55	2.35	5.49	7.15	0.66
MEDA	4.51	4.91	0.73	3.32	0.14	0.01
ACP	8.60	4.17	1.64	2.07	0.39	0.01
Latin America	17.99	2.87	15.58	11.51	2.60	0.02
Australia + New Zealand	3.06	0.96	3.28	0.53	3.81	0.06
World	59.34	55.65	52.25	59.37	40.36	2.97

NB: Candidate Countries encompass all 13 European countries that have applied to become EU members. ASEAN: Association of South East Asian Nations. 'Other Asia' encompasses South Asia, South Korea, China and Hong Kong. ACP: Africa Caribbean and Pacific and MEDA: Mediterranean Countries (see below).

Sources: For the EU: EUROSTAT/COMEXT; For the US and Japan: UN – COMTRADE.

There are a number of striking differences between the agricultural trade patterns of the triad. First, agricultural trade for Japan displays a large deficit, a situation in contrast with that of the USA and of the EU where imports are more or less of the same magnitude as exports. Second and not surprisingly, European candidate countries are only minor trade partners for both the USA and Japan, contrasting with their importance to the EU. Although geographical proximity explains the reliance of Japan on imports from neighboring Asian countries (representing more than a quarter of its imports), it does not provide a justification for Latin American food consignments in the three countries. For both the EU and the USA, Latin America is the major import source of agricultural products as well as their main trade partner. Another distinguishing feature between the EU and US trade structure of agricultural products is the greater diversification of EU import sources and export destinations. As can be seen, a policy of trade cooperation with a number of developing countries in the world has meant that ACP countries have become major importers of food products from the EU. The same applies to Mediterranean countries as well as, albeit to a lesser extent, to the ASEAN economies.

Finally, the EU and the USA are mutually important agricultural trade partners; the USA is the second most important agricultural trade partner of the EU, after Latin America, and *vice versa*. The USA exhibits a strong export specialization in oilseeds and oleaginous fruits, edible fruits and nuts, tobacco and cereals. However, the importance of trade flows between these two areas has not prevented trade disputes arising on many specific issues, such as in the case of pasta, bananas and citrus fruit in the past (see Box 1). More recently, hormones in beef, genetically modified foods, and foot and mouth contaminated sheep meat have taken preeminence. In general, the existence of mutual interests between the USA and the EU has allowed a smooth resolution of trade conflicts.

Box 1 The 'Banana dispute'

In 1993, a GATT panel was established to investigate Latin American complaints concerning the European banana trading scheme. During the years 1982–92, EC imports of bananas from ACP countries doubled, reaching 0.7 million tons. Special arrangements were made to preserve the ACP's traditional markets in the UK, France and elsewhere in the EU. In 1993, the EC Commission was proposing to impose a quota on non-ACP banana imports. According to this new regime commencing in July 1993, two million tons of non-ACP bananas could enter the EU, with increasing duties per ton, after the limit was reached; this was the equivalent of an effective tariff of 170 per cent. Given that ACP bananas are in competition, on the EU market, with exports from Central and South America, where large plantations – under control of American firms – have a cost advantage, the EU proposal was challenged by the USA (as well as by four major multinational banana producers) before the WTO. In 1997, the WTO Disputes Panel issued an interim report, ruling against the size of ACP quotas, and in favor of a 30 per cent guarantee of the EU market. However, the report accepts the principle of preferential tariffs for ACP banana exports and the exemption of Lomé from WTO rules. In this particular case, the problem for the EU was to reconcile the Union's treaty obligations under the Lomé Conventions and the obligations of the EU under the GATT.

Table 7.5 displays the shift of the EU agricultural trade balance from deficits with the USA in the mid-1990s to surpluses towards the end of the decade. In 1998, the EU still had a trade deficit with the US of €190 million. In 2001, it registered a comfortable trade surplus of €2.63

Table 7.5 Evolution of EU agricultural trade flows (1998–2001) (in billion €)

	1998		1999		2000		2001	
	M	X	M	X	M	X	M	X
USA	7.75	7.56	6.95	8.52	7.86	9.96	7.66	10.29
Japan	0.10	3.35	0.11	3.39	0.14	3.90	0.16	3.92
Candidate countries	4.97	6.04	5.35	5.45	5.59	6.34	6.49	6.70
ASEAN	4.21	1.40	4.06	1.62	4.26	2.18	4.18	2.01
MEDA	3.93	4.89	4.26	4.60	4.17	5.48	4.51	4.91
ACP	8.60	3.34	8.37	3.25	8.13	3.76	8.60	4.17
Latin America	15.07	2.74	14.84	2.41	16.62	3.03	17.99	2.87
Australia + New Zealand	2.23	0.70	2.45	0.82	2.63	0.99	3.06	0.96
World	52.46	48.71	51.84	48.06	55.86	54.44	59.34	55.65

Sources: For the EU: EUROSTAT/COMEXT; For the US and Japan: UN – COMTRADE.

billion. The candidate countries, Japan and the Mediterranean countries are other regions of the world in respect of which the EU has been recording a substantial agricultural trade surplus over the period.

Judging by the magnitude of trade flows, the major agricultural trade partners of the EU are, with the exception of Latin America and the USA, countries or regions that have enjoyed a long or intense trade partnership with the EU. These trade relationships have been governed for some time by several types of association agreements. A number of trade relationships are explored in the next section, starting with the groups of countries that have benefited most from trade agreements with the EU, that is its ex-colonies.

7.2 EU trade agreements with a number of selected developing countries

The founding principle of economic integration in the EU is free and undistorted trade between EU members. As a result, the EU has developed a full-fledged common commercial policy with the common external tariff as its core, entailing a geographically limited form of free trade. From its inception, the EU represented a special case of internal free trade and external protection. In the recent past, the EU has multiplied the number of trade agreements with third countries, and it has changed the nature and the philosophy of the agreements it has signed since the 1970s. As a consequence, the EC is involved today in a large array of agreements with most countries and regions of the world. The nature and intensity of these agreements span from the Europe Agreements signed with the CEECs to most-favored nation (MFN) treatment with non-Europe OECD countries generally, countries that receive in fact least favored access to the EU market (Box 2).

Indeed, imports from Australia, Canada, Japan, the Republic of Korea, New Zealand, Singapore and the USA, as well as those from China and Taiwan, are subject to MFN treatment in all sectors. With regard to most developing countries, the EU has developed a 'market access' strategy. Consequently, trade intensity between the EU and other countries that are structurally different, such as the LDCs, is greatly explained by these trade agreements. Traditionally confined to a list of ex-colonies, preferential access on the EU market has been granted to an increasingly large number of countries over the last few years, such as the Mediterranean countries. As was seen in Chapter 6, the former communist countries of Europe became important trade partners for EU countries in the 1990s, and have signed a number of Europe agreements paving the way to

Box 2 The Graduation of agreements involving the EC

1. Europe Agreements and Association Agreements (with 10 Central and East European Countries, Cyprus and Malta) (see Chapter 6),
2. Euro-Mediterranean agreements (involving Algeria, Morocco, Tunisia, Egypt, Jordan, Lebanon, Syria, Israel, West Bank, Gaza Strip and Turkey),
3. Free Trade Area agreements (with countries such as Iceland, Liechtenstein, Norway, Switzerland, but also with Mexico and South Africa),
4. Associations with overseas countries and territories (such as French Polynesia, and so on),
5. ACP (with 76 countries from Africa, the Carribean and Pacific Region, based on historical links; see endnote 5),
6. GSP (generalized system of preferences) with newly industrialized, middle-income and poor countries,
7. MFN (most favored nation) with developed countries, as well as with Taiwan.

membership. Finally, developing Asian and South American countries, which are geographically more remote, have experienced traditionally the least preferential treatment in their trade relationship with the EU. They normally trade under the generalized system of preferences (GSP), which implies reduced tariff rates depending on the sensitivity of products for the EU. This trade regime could be particularly helpful for developing countries in their early stage of industrialization, enabling them to use the infant industry argument. This section will look at these different trade relations and trading systems linking the EU to a selected number of developing countries.

7.2.1 Agricultural trade with Africa Caribbean and Pacific (ACP) countries

For more than two decades, sub-Saharan Africa, Caribbean and Pacific countries enjoyed the greatest preferential trade regime of all countries in the world. This is explained by the 'special' relationship that had linked these countries to some European states in the past. Indeed, the beginnings of European integration coincided with the end of the colonial period: the ex-French colonies, such as Cameroon, Mali and Senegal, became independent nations between 1958 and 1962. In 1960,

Congo, Burundi and Rwanda became independent from Belgium, whereas Somalia gained independence in the same year. With the exception of Guinea, which made a spectacular break with France, all these newly independent countries were anxious to preserve the benefits of their former association with Europe, such as a privileged access to European markets and financial aid. On the other hand, when the Common Market was formed in 1958, countries such as France, and to a lesser extent Belgium and The Netherlands, were keen to allow their ex-colonies continuing protected access onto the market of their mother countries, given the reciprocal trade gains that these countries had derived under the colonial regimes. It was therefore under French pressure that the EC took over the responsibility for allowing easy access to the producers of the former colonies of black Africa to the EC market (Molle, 1990). As a result, the first convention linking these countries with the EC, known as Yaoundé I, was signed in 1963. This convention stemmed from an earlier trade preferential system granted in 1958, involving 18 overseas countries known as the Associated States of Africa and Madagascar. When the UK joined the EEC in 1973, the scheme was extended to the former British colonies. The present scheme signed by the 15 EU member states and 71 developing countries in Cotonou (Bénin) in June 2000 follows the two Yaoundé Conventions (of 1963 and 1969) as well as the scheme that was concluded in Lomé (Togo) in February 1975. All schemes have covered the so-called ACP area, an area encompassing today 76 developing countries.[3] These ACP countries still have large agricultural sectors employing in many cases more than 40 per cent of the labor force. For example, agricultural employment represented well above 70 per cent of total employment in African countries such as Chad, Angola and Ghana in the early 1990s (World Bank, 2001).

Already under Lomé I (1976–80), 70 per cent of ACP agricultural exports and 99 per cent of all ACP exports could enter the EC market free of trade barriers (that is, custom duties and taxes with equivalent effect).[4] However, the principle of free and unlimited access to the EC market was not applied to agricultural products competing with CAP products, although these enjoyed preferential treatment when compared with similar products from third countries. Of particular significance to ACP countries has been, under Lomé I, the introduction of a system guaranteeing compensation for losses in export earnings, known as STABEX. This was a special funding facility made available under the European Development Fund (EDF), designed to cushion the shocks arising from losses of ACP export earnings due to decreases in agricultural prices on international markets. Although not all agricultural commodities were covered by STABEX under Lomé I, the list has been extended subsequently to include most agricultural primary products and some partially processed goods such as vegetable oils and cocoa paste. During the period 1995–2000, STABEX funding amounted to ECU 1.8 billion out of a total EDF budget of 12.967 billion (The Courier, 1996). The remaining €11 billion was used for structural adjustment policies, risk capital, emergency aid, regional cooperation, and so on.

In the late 1990s, ACP countries accounted for 13 per cent of all EU agricultural imports and 6 per cent of its exports in agricultural products (figures for 1998) (CEC, 2001). For many years, the EU has had a trade deficit in agricultural products with this part of the world, although this is more than compensated for by its trade surplus in manufactured goods. The largest trade deficits are obviously registered for tropical products such as cocoa, coffee, tea and spices, but also for (cane) sugar. The EU's comparative advantage in dairy products, and also to a lesser extent in cereals explains its important trade surpluses in these commodities over the years.

Attempting an assessment of the partnership with ACP countries with regard to agricultural development is made complex by the many economic, political and cultural variables at play. Funding under the EDF did help improve breeding methods and livestock production in some of these countries; it did bring more effective irrigation methods in Senegal, and it did enable the provision of technical assistance for example to the Tea Marketing Board in Burundi. There are indeed countless numbers of successful microprojects in the area of agricultural production, extension services and rural development, for example through the funding of roads and bridges by the EU (see Box 3).

However, many criticisms have been voiced in relation to the various Yaoundé and Lomé conventions. Funding under the EDF, particularly under STABEX was judged inappropriate; several decades of 'privileged' links with the EU have not been very helpful in that these countries did not experience much structural change and, as a result, they are still over reliant on their primary sector, the structure of which failed to diversify. Assessed in the background of their macroeconomic performance, these countries, particularly those of the African continent, have seen their population double since the end of the colonial rule, with a stagnant per capita income in constant terms. High population growth has put a large strain on food supplies and led to repeated famines in the region.

Box 3 Helping the setting-up of agri-business enterprises in ACP states

Working in the framework of the Lomé Convention and in close collaboration with the European Development Bank, the Center for the Development of Industry (CDI) located in Brussels has facilitated the establishment of joint ventures between European and ACP partners since the 1970s. Agri-business enterprises represent roughly 40 per cent of the CDI activities. The centre offers technical assistance for the successful development of projects spanning from the conception stage to production and distribution. The agri-business activities of the centre range from assistance in the area of packaging, production and processing of meat, to exports of organic food products. In 1996, the CDI was involved in more than 600 projects across ACP states. The activities of the CDI are seen as responding to the new development model promoted by the EU: one which favors the emergence of local private enterprises creating jobs and growth.

Source: Le Courier, N. 166, November–December 1997.

In trying to provide positive solutions to these shortcomings, the current agreement has been signed in June 2000 for a period of 20 years, with reviews every five years, and a financial protocol for each five-year period. Funding under the EDF and the European Investment Bank (EIB) amounts to €15.2 billion, to which another €9.9 million of previous EDF resources, which were uncommitted, should be added. Compared with the previous conventions, the Cotonou Convention adds a definite and strong political dimension. Among its ambitious goals is that of eradicating poverty in all ACP countries, and of facilitating their integration into the fast developing global economy. The main provisions of the various and successive schemes have been: tariff preference, special treatment for products coming under the CAP and development aid through technical assistance. In relation to trade, the two zones have committed themselves to conclude new WTO compatible arrangements by phasing out barriers to trade between themselves. The EU has embarked since 2000 on a process that will give free access to the bulk of products from all LDCs by 2005 at the latest. The Cotonou Agreement set out the objectives and the basic principles for Economic Partnership Agreements (EPAs) between the two regions.[5] Opened in September

2002, the EPA negotiations should be completed by 1 January 2008. Seen as complementary to ACP–EC development cooperation and as a cornerstone in the EU policy aimed at deepening economic relations with developing countries, EPAs govern development, trade as well as political aspects of EU–ACP relations. Committed to the Doha Development Agenda, in which the EU strongly supported putting development at the heart of the multilateral trading system, the EU is keen to open up its market to, eventually all imports from least developed countries (see Chapter 8).

7.2.2 EU–Mediterranean agricultural trade

The Mediterranean Region, comprising Algeria, Morocco, Tunisia, Egypt, Jordan, Lebanon, Syria, Israel, West Bank, Gaza Strip and Turkey represents roughly 10 per cent of the EU agricultural exports, and 7.5 per cent of its imports (CEC, 2001). In relative terms, this is more or less equivalent to the case of EU–ACP agricultural trade. The difference between ACP and Mediterranean countries is that the latter have had a trade deficit in agricultural commodities with the EU for many years. Also, this region has been one of the least dynamic regions in terms of export growth over the last 15 years. The weak export performance has taken place despite a well-established economic co-operation with the EU since the early 1970s with the signing of bilateral trade agreements. For more than 20 years, Mediterranean agricultural exports in which the EU displayed low self-sufficiency rates – such as avocados – were accorded limited preferential access on the EU market, while Mediterranean industrial goods enjoyed duty-free access to the EU market. Here again, sensitive agricultural products were excluded from the agreements. In addition, the EU provided financial resources, through EIB loans, to help these countries transform structurally. A certain degree of structural change has taken place in some of these countries in that their trade structure has shifted from a strong dominance of agricultural and energy products towards a predominance of manufactures over the last 25 years. Although this group of countries is still the EU's most important trading partner from the developing world, its importance has been dwindling during the last decade or so. There are two major economic reasons that explain the erosion of preferential and exclusive ties that this group of countries had with the EU:

- First, the recentering of the EU's external policy in favor of the Central and East European countries, as well as its awakening to the South East Asian region, to China and to other parts of Asia.

- Second, the ongoing and generalized process of trade liberalization at the global level, in particular for agricultural products.

In order to remedy this situation, the response of the EU has been in the mid-1990s to try consolidate the partnership with these countries through a revision of the old style EU–Mediterranean agreements and the signing of a new deal known as the *Barcelona Agreement.* Before we venture into some of the details of this new agreement, we will briefly review the principal characteristics of these countries' agricultural sectors and of the EU–Mediterranean trade in agricultural products.

7.2.2.1 The agricultural sector in the Mediterranean countries and EU–MED agricultural trade

Although the EU's largest imports from the Mediterranean countries are petroleum and clothing products, the agricultural sector still plays an important role as a generator of foreign exchange in most of the countries in the region. In the mid-1990s, agricultural and food products represented 22 per cent of total exports to the EU for Morocco, 17 per cent for Turkey, 16 per cent for Lebanon and more than 10 per cent for Tunisia, Israel and the occupied territories (see Table 7.6).

The relative importance of agricultural exports for these countries can be gauged with the use of a revealed comparative advantage indicator (RCA). An RCA indicator gives an indication of the relative importance of a product in total trade of a given country, compared with a reference

Table 7.6 MED 12 exports of agricultural and food products to the EU as a percentage of total exports to the EU (1995) (in %)

Algeria	0
Tunisia	11
Morocco	22
Egypt	9
Syria	2
Lebanon	16
Jordan	8
Israel/West Bank – Gaza Strip	13
Malta	1
Cyprus	26
Turkey	17

Source: CEC (1997a).

group of countries. For any country *j*, and any given product *i*, the *RCA* can be defined as:

$$RCA = \frac{x_{ij}/X_j}{\sum_j x_{ij}/\sum_j X_j}$$

with x_{ij} = exports of product *i* from country *j*, and X_j = total exports of country *j*.

We have calculated the *RCA* index in agricultural products for a number of countries in 1997, as shown in Table 7.7. For our purpose, we have taken the group of low-to-middle income countries in the world as a comparable group of reference. It goes without saying that, had we taken Europe and Central Asia as a reference group, the resulting indices would all be greater than those shown in Table 7.7. The results in the table imply that Morocco, Jordan, Syria and Turkey all have an index well above the 1 threshold, implying a strong export specialization in agricultural products. On the other hand, Algeria, Tunisia, Egypt and Israel do not hold an RCA in food products. These countries have moved away from the agricultural sector and along the developmental path. In particular, Algeria, Tunisia and Egypt exhibit a heavy dependence on exports of labor-intensive commodities, such as textiles and clothing, as well as on mineral fuels, such as crude oil and gas (in the case of Algeria). Finally, Israel is the only country of the group that has been able

Table 7.7 Revealed comparative advantage of Mediterranean countries in agricultural products

	Food products*	**Fruits and vegetables#**
Algeria	0.026	0.3
Morocco	2.359	8.1
Tunisia	0.760	1.1
Jordan	1.536	5.4
Egypt	0.196	2.6
Lebanon	na	9.3
Syria	1.367	2.2
Israel	0.309	2.3
Turkey	1.130	6.6

Sources: * Author's calculations based on World Development Indicators (World Bank, 2001); # CEC (1997a) European Economy.

to diversify away from labor-intensive industries towards higher VA manufacturing commodities, such as electrical and electronic products. In a way, Table 7.6 reveals also the large structural heterogeneity of Mediterranean countries.

The significance of agricultural products in some of the Mediterranean countries' exports reflects the relative importance of their agricultural sector in their productive structure. In Morocco, Syria, Jordan and Turkey, agricultural production still accounts for 15 to 25 per cent of GDP and for 23 to 45 per cent of employment (World Bank, 2001). The table also shows a strong specialization in some of the agricultural commodities. Fresh fruits and vegetables, but also fish and seafood, are among the main agricultural exports of these countries. These products are in direct competition with South European agricultural production and exports; Italy, Spain, Portugal and Greece are also large producers of fresh fruits, tomatoes, olive oil and citrus products.

As was seen earlier, the EU is the main export destination for the Mediterranean countries' commodities and food products. For example, in the mid-1990s, Morocco and Syria exported around 60 per cent of their agricultural and manufacturing output to the EU (CEC, 1997a). Jordan is the only country in the group displaying a very low export dependence on the EU market. Fruits and nuts, vegetables, and floricultural products are the main EU imports from these countries. These three commodities accounted for nearly 70 per cent of all EU agricultural imports from this region in 1998 (CEC, 2001). The EU agricultural exports to these countries are more diversified. Sugar (13.7 per cent of the total), dairy products (13.5 per cent) and cereals (13 per cent), were the major products exported by the EU in 1998. Again, the geographical proximity to the EU has prompted these countries to sign many bilateral agreements with the EU. The principal characteristics of EU–MED trade relations under the 'old style' Mediterranean agreements can be summarized as follows:

(i) a strong asymmetry with regard to market access (exports from the EU receive most favored nation treatment), and a strong dependence of the Mediterranean countries on EU markets. The EU is a major export destination for Mediterranean products, whereas these countries are only marginal trade partners for the EU;
(ii) very limited intra-regional trade, and insufficient export diversification in the Mediterranean countries;
(iii) continuing trade surplus in favor of the EU. In 1998, the EU agricultural imports from the Mediterranean countries represented less than 78 per cent of its agricultural exports to these countries.

The characteristics of EU–Mediterranean trade relations as summarized above, combined with many political elements have prompted a thorough revision of the EU agreements with these countries in recent years. This is known as the *Barcelona Declaration* and its ensuing *Barcelona Process.*

7.2.2.2 The Barcelona Agreement and its significance for agricultural trade

The Barcelona Act of November 1995 is seen as an important step in the process of *rapprochement* between the EU-15 and 11 South Mediterranean countries including the Palestinian Authority (West Bank and Gaza Strip) as well as Mauritania. The Barcelona Agreement takes the form of 'new style' EU–MED agreements building on those signed for example with Tunisia and Israel in the past. For example, the 1964 trade agreement signed by the EC and the State of Israel was aimed at promoting trade in manufactures and in agricultural products between the two parties. It envisaged a number of reduced custom duty rates for products such as citrus fruits, avocados, dates, bananas and nuts, all products originating in Israel (OJEC, 1974). The 'new style' agreements are followed by agreements between members themselves. Aimed at establishing a common Euro-Mediterranean area of peace, stability and shared prosperity, the agreement allows for a gradual setting up of a Free Trade Area between the EU-15 and the 11 Mediterranean countries by 2010. The ultimate goal of the new partnership is to improve the trade performance of the Mediterranean countries through trade liberalization and modernization, intra-regional trade, thereby creating a more favorable investment climate, promoting growth with employment.

The Barcelona declaration covers three mutually reinforcing areas that have become the three cornerstones of EU external policy in recent years (see Section 7.3); these encompass a political and security partnership, an economic and financial partnership and a social, cultural and human partnership. The EU–Mediterranean partnership is therefore conditioned by the full acceptance of the trade liberalization clauses by the Mediterranean countries, and also by their acceptance of the political, social, cultural and human aspects of the Barcelona declaration. On the economic front, the argument is that reciprocal trade liberalization will impose strong pressures on industries of the Mediterranean countries, leading to inevitable structural change. This will in turn facilitate the diversification of exports for these countries. Structural change is facilitated by increased financial assistance of the EU, under the form of project aid, budgetary aid and technical assistance.[6]

In the agricultural sector, the Barcelona declaration envisages that trade in agricultural products be 'progressively' liberalized, through preferential and reciprocal access among parties. Agricultural trade liberalization would be extremely gradual, implying for the time being a rather weak trade liberalization of agricultural and fisheries products. In particular, the new style EU–MED countries' agreements allow tariff concessions to products such as fruits and vegetables, durum wheat, olive oil and spices, in many cases with quantitative limits, given that many of these products are in direct competition with EU production. Commodities such as meat, dairy products and cereals face MFN tariffs on the EU market, which corresponds to the least favorable treatment. However, with full and immediate liberalization, the agricultural sector of Mediterranean countries would be positively affected by this new *rapprochement,* for their agricultural products would be allowed wider access on the EU market, an objective that complies with the WTO requirements (see Chapter 8). According to Garcia Alvarez Coque (2002), the full liberalization of agriculture is a *sine qua non* for the reform process to generate the substantial benefits as envisaged by the Barcelona Declaration. As estimated by Lorca and Escribano (2000), full agricultural liberalization would lead to substantial export gains, from 0.4 per cent of GDP for Tunisia, to 3.3 per cent for Egypt, over a five-year period.

Regular follow-up meetings of Euro–Mediterranean ministers for trade have been held since May 2001.[7] These meetings allow, *inter alia,* to review the progress made in achieving the stated objectives. For example, concrete results in the field of rules of origin have been obtained quite rapidly. Nevertheless, the new style EU–MED agreements do not seem to favor the transformation and processing of agricultural products by the Mediterranean countries, as most of the tariff concession is granted only for the manufacturing component of the products (Garcia Alvarez Coque, 2002). As suggested by this author, the enlargement to the CEECs may lessen the opposition to including the agricultural 'sensitive products' in the free trade provisions of the Barcelona Agreement.

7.2.3 Agricultural trade links with Asia

Asia, like Latin America, has had its trade with the EU governed by the GSP system since 1971. Aimed at permitting the integration of developing countries into the world economy and into the multilateral trading system, GSP is based upon the principle of duty-free entry for nearly all manufactured products and reduced rates of duty for selected agricultural and processed agricultural products (CEC, 1995). Since its inception, the GSP system operated by the EU has been revised a number of times. For example, through its revision in 1997, the new scheme grants preferential access to EU markets to a wide range of agricultural commodities from the developing world, except for those products that are subject to a market organization (CEC, 2001: 167).[8] Also, the 1997 revision of the GSP system provided specific arrangements for countries undertaking to abide by social and environmental standards.

In contrast with the ACP and Mediterranean countries, Asian countries in general have experienced a high degree of economic dynamism since the 1970s. An uninterrupted cycle of economic boom, only marginally jeopardized by the 1997 economic crisis, has transformed these countries dramatically, both from an economic and a political point of view. In spite of its efforts, Asia as a region is still nevertheless home to two-thirds of the world's poor, with 800 million people living on less than $1 per day. Questions of food security, health and access to basic services are therefore still pressing issues in the lower-income countries of the region. Understandably, Asia is a very diverse region, economically, politically, socially and culturally. The region includes the two most populous countries in the world (China and India), and some of the smallest states such as Brunei and Bhutan; it encompasses some of the richest countries in the world (Japan and Singapore), and some of the poorest (eight Asian countries, such as Myanmar are on the UN least developed list). The agricultural sector is still the major absorber of surplus labor in many of the Asian countries, employing respectively 66.6 and 63.2 per cent of the total labor force in countries such as Myanmar and Bangladesh in 1996. In contrast, the agricultural sector in Japan is minuscule, accounting for less than 0.5 per cent of total Japanese merchandise exports in 1999 (World Bank, 2001).

Amongst all Asian countries, South-Eastern and East-Asian countries have received most of the attention of the EU in the recent past. Although there is an EU–India partnership, and although there are attempts at forging links with countries such as Pakistan and Mongolia, it can be stated that South Asia in general, and more notably Central Asia, have been relatively neglected. In particular, the establishment of ASEAN (Association of South-East Asian Nations) in 1967 has been an important catalyst for the future development of EC–Asian economic and diplomatic relations. This association was formed between Indonesia, Malaysia, the Philippines, Singapore and Thailand. Brunei joined in early 1984 after achieving independence, Vietnam in July 1995, Myanmar and Laos in July 1997, and Cambodia was the last member to join. The EC–ASEAN dialogue of the 1970s culminated with the EC–ASEAN Co-operation Agreement signed in 1980,[9] and it provided a basis for the 'new Asia strategy' endorsed by the 1994 Essen European Council. Calling for a higher profile of the EU in Asia (CEC, 1994), this 'new Asia strategy' led to the first ASEM summit in 1996 (Asia–Europe Meetings),[10] as well as to the new 'strategy for enhanced partnership' with Asia, adopted by the Commission in its 2001 communication and covering the years 2001–10 (CEC, 2001). The key objective of this strategy is to strengthen the presence of the EU in Asia, by focusing on several dimensions, amongst which are the improvement of market access for the poorest developing countries and the reduction of poverty. Also, a greater importance is accorded to South Asian and North-East Asian countries in this strategy.

It can be seen that China, which restored its diplomatic relations with the EEC in 1975, is now an integral component of the new EC policy towards Asia. This is explained, on the EU side, by the enormous potential that the Chinese market represents for EU traders and investors alike. Indeed, the new policy is the EU's response to the changing political and economic situation in the region as a whole. Asia has become a crucial economic partner for the EU. The region as a whole is the EU's third largest regional trading partner, and it accounted for 21 per cent of total EU external exports in 2000, after Europe outside the EU (31 per cent) and NAFTA (28 per cent) (EUROSTAT, 2001).

As trade cooperation played an important role in the 1980 EC–ASEAN Cooperation Agreement, the opening to Western and in particular EU markets, helped change the pattern of trade for these countries. In the case of the ASEAN countries, the share of primary and agricultural products has decreased substantially, reaching for example 3.02 per cent of Thai merchandise exports in 1999 compared with 11.24 per cent in 1980 (World Bank, 2001). The EU imports of food products from ASEAN countries have oscillated around 7 per cent during the 1990s. Agricultural imports from Japan – one of the most economically advanced countries in the region – have also stagnated at around 0.2 per cent of EU total agricultural imports (Table 7.8).

On the other hand, South Asian countries such as India have increased their share of agricultural products on the EU market during the 1990s. These countries are less developed than the oldest members of the ASEAN (such as Thailand and the Philippines), and they still have a relatively large agricultural sector. Increased market access provided by the EU has favored the increased penetration of these countries' agricultural products on the EU market. The growing importance of the EU market as an important outlet for India's agricultural products should be

Table 7.8 EU–Asia agricultural trade (1999)

	EU imports (as a % of total)	**EU exports (as a % of total)**
SAARC*	2.6	1.1
India's share	2.0	0.5
China	3.0	1.5
Japan	0.2	7.2
ASEAN	7.5	3.4
Extra EU–15	100.0	100.0

NB: * SAARC: South Asian Association for Regional Cooperation. This comprises India, Pakistan, Bangladesh, Maldives, Sri Lanka, Nepal and Bhutan.
Source: CEC (2002a).

highlighted. In the year 2000 alone, India's agricultural exports to the EU increased by 10.5 per cent (EUROSTAT, 2001). These developments have contributed to increasing the trade surplus that these countries have enjoyed with the EU in recent years.

The complementarity between EU and Asian agricultural products leaves more scope for further trade increases between the two regions.

7.3 Tackling rural poverty and malnutrition in the developing world: EU cooperation with LDCs

In the developing world, poverty afflicts mostly, yet not exclusively, those engaged in the agricultural sector and those living from non-farm activities in rural areas (IFAD, 2001). In recent years, the EU has endeavored to assert its role in fighting poverty through its development policy with developing countries. The EU development strategy with regard to LDCs follows a broad and holistic framework that was designed in the background of the many failed and piecemeal polices of the past. It is now recognized that tackling one issue at a time is counterproductive, and that the reduction of rural poverty in the developing world must involve the selection of policies addressing the entire rural space, encompassing many aspects ranging from food security and technology transfer to good governance issues. This all-encompassing strategy echoes in some way the revision of the definition of poverty provided in recent years. Following the work of Nobel Price A. Sen, poverty cannot simply be defined in terms of per capita incomes, but it involves the inclusion of many other indicators such as access to health, education and clean water. These are in turn connected with societal and political aspects of economic development. Before we examine the role the EU aspires to play in developing countries in terms of poverty reduction through policies targeted at the agricultural sector, we will provide first a broad indication of poverty in LDCs.

7.3.1 Overview of poverty in the developing world

According to the FAO (2002), undernourishment affected some 840 million people in 1998–2000, most of which (95 per cent) live in LDCs. Although some countries such as China, Vietnam and Peru, have managed to reduce the number of hungry people since 1990–92, the total number of undernourished people has in fact increased by more than 50 million in the last decade, if we were to set the figures for China aside. More than 2 billion people in the world suffer from micronutrient malnutrition explained by the lack of vitamins or minerals; for example it has been estimated that between 100 and 140 million children worldwide are exposed to vitamin A deficiency leading to blindness. The main cause of hunger is poverty; drought and armed conflicts are other explanations. These staggering figures call for immediate actions, if, as the FAO report adds, the goal set at the 1996 World Food Summit – that is to halve the number of hungry people by 2015 – is to be taken seriously.

World hunger and malnutrition represent enormous costs, not only for the people affected by this calamity, but also for their communities and for their entire countries. The FAO (2002) estimates that the global benefits of reducing the number of hungry people by half is at least US$120 billion per year. The world organization calls for an additional yearly public investment of US$24 billion to reach the target set in 1996. Bringing a solution to this vital problem is therefore in the interest of the entire world community. Given that the EU is a major global player on world food markets, what can be its role in helping poor countries address this issue?

7.3.2 The EU development and aid policy

In addition to being an important trade partner for many less developed countries, the EU has also become a major player in their development process over the years. Covering all regions of the world, the EU provides nearly half of all public aid to developing countries. Based on the concept of solidarity, the Community Development Policy promotes sustainable, equitable and participatory human and social development.[11] The objectives of the EU international development policy are included in the Treaty establishing the Community (Title XX, Article

177). These objectives are to reduce, and eventually, to eradicate poverty in the developing world, through the support for economic, social and environmental development. Poverty is defined here in its broader sense, including the notion of vulnerability and denied access to education, health, drinking water, land, employment, credit facilities, information and political involvement. In its recent communication to the Council pertaining to the EC Development Policy, the Commission has drawn up six areas of action, of which are food security and sustainable rural development (CEC, 2000). It should however be noted that the Community sees sustained growth as a prerequisite for poverty reduction. Much of the EU developmental action is to help the LDCs become inserted in the world trading system, providing them the opportunity to become increasingly competitive at both the regional and world levels. This explains why the EU has devoted much effort in concluding a large number of trade agreements with many countries of the developing world. In addition, the EC Development Policy fosters two other types of actions: food emergencies and development assistance. The EDF is the main channel through which financial transfers provided under the EU food aid initiative and development assistance policy can reach the developing countries. In particular, the EU and its member countries are signatories to the Food Aid Convention, 'the aim of which is to contribute to world food security and improve the national community's ability to respond to food emergencies and other food needs to developing countries' (CEC, 2000: 168). Examples of countries benefiting recently from food aid emergency include Zimbabwe, Malawi and Zambia in October 2002. Food and humanitarian aid was increased to €310 million to help these countries fight rampant malnutrition, disease and the risk of famine. Some €177 million were managed directly by the EU, whereas the aid was delivered via the World Food Program,[12] non-governmental organizations and governments. The creation in January 2001 by the EC Commission of Europe Aid, a cooperation office for the management of EU external aid in favor of developing countries is aimed at accelerating the implementation and at improving the quality of aid projects in the world. Europe Aid is responsible for 80 per cent of total EU external aid, representing more than €9 billion per year, attributed to development projects in poor countries.

The agricultural and agri-food sectors have been large beneficiaries of EU aid financial transfers so far. For example, in Asia alone, the EU as a whole (that is the EC Commission plus the individual member states) accounts for 30 per cent of total aid flows to developing Asia. EU aid developmental projects in this part of the world have been implemented with the following aims:

- to help some 32,000 rice farmers in Bangladesh to give up pesticide-intensive farming in favor of environmentally sound production methods,
- to help rural Cambodian families to grow enough food to meet their own needs,
- to help the fishing fleet in China's Fujian province to increase its productivity and safety,
- to establish self-sustainable village organizations in India's Gujarat State,
- to resettle 8000 farm families in Sri Lanka,
- to re-integrate Vietnamese boat people who have chosen to return to their homes, and
- to improve basic agricultural education in the Philippines.

These are some of the 268 development projects, which the EU was financing under its program of development assistance to Asia in the mid-1990s. The total expenditure on these projects amounted to ECU 5.2 billion, of which the EU's contribution was ECU 2.1 billion. (Between 1976 and 1995, the EU devoted some ECU 2.9 billion on completed and ongoing development projects.) In this region of the world, India has traditionally been the main beneficiary of the funds provided by the EU for development assistance and economic cooperation. As of 30 June 1995, the total amount of the projects funded in India by the EU came to ECU 740 million. The other major beneficiaries were: Bangladesh (ECU 342 million); Philippines (ECU 190 million); Pakistan (ECU 181 million); Indonesia (ECU 170 million); Thailand (ECU 127 million); Vietnam (ECU 88 million): Cambodia (ECU 76 million); China (ECU 65 million); and Sri Lanka (ECU 45 million) (CEC, 1996a).

Cooperation in the agri-food sector has been stimulated by the establishment of the European Community Investment Partners Scheme (ECIP) in 1988. Fruit of the Cheysson facility, the focus of the scheme is to encourage cooperation between SMEs, through grants and interest free loans. The scheme supports the different phases in the creation of joint ventures in eligible developing countries of Asia, Latin America and the Mediterranean and South African regions. Although most of the projects approved under ECIP were manufacturing projects (for example 57 per cent of all projects in Asia between 1988 and 1997), the agricultural and agri-food sector have attracted more than 10 per cent of the funding under this scheme. This scheme presents mutual benefits to both partner countries: it is of great benefit to European putative small firms who wish to set up a foothold in developing countries; the benefit to the host economies is directly measurable through the number of jobs created by the scheme; between 1988 and 1996, an estimated 30,000 new jobs have been created in the eligible countries.

Finally, as was discussed in Chapter 6, the emphasis in the last decade has been on agricultural research for development, and also on collaborative research with the purpose of helping the LDCs. For example, EIARD (European Initiative for Agricultural Research for Development), is designed to foster European research and technology to the benefit of developing countries, including the Mediterranean partner countries, in all sectors related to agriculture. The aim of this initiative is to achieve a second 'double green' revolution, that is one that brings many more beneficial impacts to the different actors of the economic system, in both developed and developing countries, and that is compatible with the objective of sustainable development. At a broader level, the EU is a donor to the CGIAR (Consultative Group on International Agricultural Research), and also to the developing countries' NARS (National Agricultural Research Systems). The CGIAR is supported by some 45 donor countries and organizations, and co-sponsored by a number of international organizations, including the FAO and the World Bank. In 1996, Europe's contribution to the budget (US$300 million) was the largest (42 per cent), compared with 15 per cent each in the case of North America and the Pacific Rim (Japan and Australia) (CEC, 1997b). Although it may be argued that Europe's lead is weakened by its lack of concerted position at CGIAR meetings, this is another instance which shows clearly the commitment of the EU for the improvement of economic and social welfare in the poorest countries of the world.

Summary

Although the share of agricultural products in world trade has fallen from around 20 per cent to 10 per cent between 1970 and 1999 (UNCTAD, 2002), exports of food commodities are still an important source of foreign exchange for many countries of the world. In particular, much of the EU trade involving developing countries, in particular African countries, is still dominated by primary commodities and by agricultural products. The 1990s have nevertheless born witness to two unprecedented phenomena: first the composition of world agricultural trade has changed geographically, with an increased participation of developing economies, and growing intra-developing countries trade. Second, the structure of world demand for agricultural products has shifted to such an extent that some authors have talked about a 'livestock revolution' in the LDCs; in the Northern hemisphere, demand has shifted towards tropical products.

The diminishing influence of the EU as an agricultural trade partner in recent years has been compensated for by its willingness to assert itself more fully on the international developmental scene. Through the signing of trade agreements with most countries of the developing world, the EU is aiming at becoming a major actor in terms of economic development; it aspires at playing a greater role in reducing poverty and malnutrition throughout the world. The 'new style' agreements signed with ACP countries (Cotonou Convention), and with the Mediterranean countries (through the so-called Barcelona Process), mirror the new partnership philosophy underlying the EU agreements with other parts of the world; these agreements place a large emphasis on political and social aspects, for example by situating social development at the heart of the developmental process. However, a major criticism that needs to be addressed by the EU is its still too pronounced propensity to exclude its agricultural 'sensitive' products from the agreements.

The increased participation of developing countries in world agricultural trade during the 1990s has however been a very uneven phenomenon. The involvement of the 49 least developed countries in world trade, and in agricultural world trade, has been eroded. Their total export share fell from 3 per cent in the 1950s to around 0.5 per cent since the early 1980s (UNCTAD, 2001). Moreover, exports of African agroindustrial products in the world market have plummeted during the past several decades (Diaz-Bonilla and Reca, 2000). Aware of the risks of potential uneven development arising from trade liberalization, the EU is now eager to allow greater market access to these countries. Its position has indeed been vehemently asserted at the recent Doha Conference initiating a new round of multilateral trade negotiations (CEC, 2002b).

Key terms and concepts

Comparative advantage
Revealed comparative advantage
'Sensitive' (agricultural) products
Trade agreements

Most favored nation
Generalized system of preferences
Preferential market access
Poverty eradication
EU development policy
Food and humanitarian aid

Notes

1 In order to measure the production costs, Ricardo uses the labour theory of value, which has been central to the development of Marxist theory. As a consequence, Gottfried Habberler introduced the notion of opportunity cost as a means of discarding the labour theory of value, without nevertheless jeopardizing the law of comparative advantage.
2 Intra-industry trade refers to trade within the same industry (that is at a detailed level of analysis, EU beef trade against Argentinean beef). The increase of intra-industry trade has been paralleled with the increase in income levels in regions such as the EU, leading to the increase in consumer sophistication, and to ensuing product differentiation effects.
3 ACP states that have joined in recent years include South Africa and Cuba. Note that South Africa has also concluded a Trade and Development Co-operation Agreement (TDCA) with the EU. Although Cuba is also an ACP member, it was not a signatory to the Cotonou Agreement. East Timor has applied for ACP membership.
4 Agricultural products are not the major source of export earnings for these countries, for up to 60 per cent of ACP exports to the EU today are raw materials, which enter the EU market free of duty.
5 The 76 ACP countries eligible for EPAs are: Angola, Antigua and Barbuda, Bahamas, Barbados, Belize, Benin, Botswana, Burkina Faso, Burundi, Cameroon, Cape Verde, Central African Republic, Chad, Comoros, Congo (Brazzaville), Congo (Kinshasa), Cook Islands, Côte d'Ivoire, Djibouti, Dominica, Dominican Republic, Equatorial Guinea, Eritrea, Ethiopia, Fiji, Gabon, Gambia, Ghana, Grenada, Guinea, Guinea-Bissau, Guyana, Haiti, Jamaica, Kenya, Kiribati, Lesotho, Liberia, Madagascar, Malawi, Mali, Marshall Islands, Mauritania, Mauritius, Federal States of Micronesia, Mozambique, Namibia, Nauru, Niger, Nigeria, Palau, Papua New Guinea, Rwanda, Saint Kitts and Nevis, Saint Lucia, Saint Vincent and the Grenadines, Samoa, Sao Tome and Principe, Senegal, Seychelles, Sierra Leone, Solomon Islands, Somalia, Sudan, Suriname, Swaziland, Tanzania, Togo, Tonga, Trinidad and Tobago, Tuvalu, Uganda, Vanuatu, Wallis and Futuna, Zambia, Zimbabwe.
6 With grants from the MEDA program, and with substantial loans from the EIB, total funding made available by the EU to the Mediterranean partners was 2bn Euros in each year of 1998 and 1999.
7 The Second Euro-Mediterranean Ministerial Conference on Trade was held in Toledo (Spain) on the 19th March 2002.
8 Products are classified into several categories, depending on their sensitivity. For example 'non-sensitive' products are entirely free of custom duty, whereas

the highly sensitive, sensitive or semi-sensitive products are subject to a tariff (determined by the EU and accepted under the Uruguay Round).

9 Normally, after joining the ASEAN, the new members need to ratify a bilateral Co-operation Agreement with the EC, before the Protocol extending the Co-operation Agreement between the EC and ASEAN can be negotiated and ratified. Consequently, Vietnam signed such a Co-operation Agreement in 1995, and Lao in April 1997. Myanmar, faced with its human rights problems, did not sign a bilateral agreement with the EU prior the negotiation pertaining to the extension of the EC-ASEAN Co-operation Agreement.

10 ASEM (the Asia-Europe Meeting) is an informal process of dialogue and cooperation bringing together the fifteen EU Member States and the European Commission, with ten Asian countries (the 7 oldest members of ASEAN, Japan, South Korea, and China).

11 For an understanding of how the concept of solidarity has been central to the building of the EU and to policy making therein, see Andreosso-O'Callaghan (2002).

12 The World Food Program, the largest food aid organization in the world is the UN frontline hunger relief agency. In 2001, the WFP helped feed more than 77 million people in 82 countries.

Bibliography

Anderson, K. (1988) *Rent-seeking and Trade Policy in Rich and Poor Countries,* mimeo, University of Adelaide, September.

Andreosso-O'Callaghan, B. (2002) 'The centrality of the solidarity concept in the process of European integration', *Studies – An Irish Quarterly Review,* 91: 362, 134–43, Summer.

CEC (1994) *Towards a New Asia Strategy,* COM (94) 34 final (Brussels and Luxembourg).

CEC (1995) *Trade Relations between the EU and the Developing Countries* (EU Commission, DG for Development, Brussels and Luxembourg).

CEC (1996a) *Frontier Free Europe,* A Monthly Newsletter – Supplement, No. 4, 1996, Brussels.

CEC (1996b) *Creating a New Dynamic in EU-ASEAN Relations,* COM (96) 314 final (Brussels and Luxembourg).

CEC (1997a) *The EU as a World Trade Partner,* European Economy No. 3 (Brussels and Luxembourg).

CEC (1997b) *European Initiative for Agricultural Research for Development* (Communication to the Council and the European Parliament) COM (97) 126 final (Brussels and Luxembourg).

CEC (2000) *The EC Development Policy,* Communication to the Council and to the EP, COM (2000) 212 (01) (Brussels and Luxembourg).

CEC (2001) *EU & Asia: A Strategic Framework for Enhanced Partenership.* Communication from the Commission, COM (2001) 469 final, Brussels and Luxembourg, 4th September.

CEC (2002a) *The Agricultural Situation in the EU* (2000 Report, Brussels and Luxembourg).

CEC (2002b) *Trade and Development – Assisting Developing Countries to Benefit from Trade.* Communication from the Commission to the Council and the European Parliament, COM (2002) 513 final (Brussels and Luxembourg).
Delgado, C., Rosegrant, M., Steinfeld, H., Ehui, S. and Courbois, C. (1999) *Livestock to 2020: the next revolution,* IFPRI Discussion Paper, Washington DC.
Diaz-Bonilla, E. and Reca, L. (2000) 'Trade and agroindustrialization in developing countries: trends and policy impact', *Agricultural Economics,* 23: 3, 219–29.
EUROSTAT (2001) *Trade Statistics of the EV* (Luxembourg).
EUROSTAT (2002) *Statistical Yearbook on Candidate and South–East European Countries* (Luxembourg and Brussels).
FAO (2002) *The State of Food Insecurity in the World 2002.* Annual Report of the Food and Agricultural Organization, Rome.
Garcia Alvarez Coque, J.M. (2002) 'Agricultural trade and the Barcelona Process: Is full liberalisation possible?', *European Review of Agricultural Economics,* 29: 3, 399–422.
IFAD (2001) *Rural Poverty Report 2001 – The Challenge of Ending Eural Poverty.* International Fund for Agricultural Development, Rome.
Lorca, A. and Escribano, G. (2000) *The Euro-Mediterranean Free Trade Area. From Competition to Integration.* Paper presented at the Seventh Economic Research Forum International Conference, Amman.
Molle, W. (1990) *The Economics of European Integration – Theory, Practice, Policy* (Dartmouth: Aldershot).
OECD (2000) *Trade by Commodities* (Paris: OECD).
OJEC (1974) 'Council Decision of 8 May 1964 on the conclusion of a Trade Agreement between the EEC and the State of Israel', *Official Journal of the European Communities,* No. 95, 64/357/EEC, Brussels.
Ricardo, D. (1817) *On the Principles of Political Economy and Taxation.* Reprinted in P. Sraffa and M. Dobb (eds) (1953) *The Works and Correspondence of David Ricardo,* Vol. I (Cambridge: Cambridge University Press).
Smith, A. (1776) *An Enquiry into the Nature and Causes of the Wealth of Nations* (Reprinted 1976, Chicago: Chicago University Press).
The Courier (1996) No. 155, Europan Commission, January, Brussels.
UNCTAD (2001) *Duty and Quota-Free Market Access for LDCs: An Analysis of QUAD Initiatives.* Geneva.
UNCTAD (2002) *Trade and Development Report Report* (United Nations Conference on Trade and Development, Geneva).
World Bank (2001) *World Development Indicators,* Washington.

8
Agricultural Trade Liberalization

Objectives of this chapter

- To present briefly the salient points pertaining to agriculture in the Uruguay Round of GATT negotiations and the measurement of agricultural trade distortions.
- To highlight the negotiating position of both the EU and the USA.
- To assess very briefly the CAP–GATT link.
- To provide an analysis of the impact of trade re-ordering (*ex-ante* and *ex-post* analyses).
- To look at the future: the 2001 Doha agenda and beyond.

Introduction

As was seen in the previous chapter, agriculture is still of particular significance to the EU, as it is the major importer and the second largest exporter of food products in the world. Consequently, the EU is a very prominent and vocal member of what is today known as the World Trade Organization (WTO). Successful rounds of multilateral trade negotiations in the past have led to a decrease in the level of protection facing developing as well as developed countries. The situation of the agricultural sector, where protection in the developed countries grew enormously in relative terms until the mid-1980s, is generally in contrast with a background of falling tariffs experienced in the manufacturing sector. Indeed, average tariff levels on manufacturing products have decreased from 40 to 50 per cent of the import value in 1950 to an average of 3.9 per cent in the 1990s. Thanks to the last round of multilateral negotiations, the Uruguay Round (UR), average tariffs on fish products declined by more than a quarter (Senti and Conlan, 1998).

Whether falling tariffs have meant freer trade is debatable as current imports of agricultural products can face many other types of non-tariff barriers (NTBs), some of them on the rise. Indeed, as was seen in Chapter 3, obstacles to trade can take the form of NTBs such as, quantitative restrictions with equivalent effect, import licensing schemes, antidumping and countervailing duties, quotas, voluntary export restraints and price control measures that restrict imports by artificially increasing the price of imported products. In recent years, the importance of phytosanitary and veterinary measures has increased. The latter, which hamper trade in sectors such as beef, dairy products and tobacco, have tended to become more prominent in recent years with the development of genetically modified organisms (GMOs). From the developed economies' point of view, the violation of intellectual property rights (IPRs) – mostly by developing nations – results also in substantial trade distortions. IPRs include patents, copyrights, trademarks, trade secrets, and they are particularly of primary concern to countries with large biotechnology industries.

Inspired by the post-world war American and British vision of a new international trading system of free and democratic nations, the General Agreement on Tariff and Trade was signed in 1947 by 23 countries with the aim of promoting free trade in goods.[1] In its original form, GATT was a provisional treaty and not an organization (Senti and Conlan, 1998). The GATT has grown both in size and scope through the eight multilateral trade negotiations rounds organized since its creation: the Geneva Round (1947), the Annecy Round (1949), the Torquay Round (1950/1951), the Geneva Round (1955/56), the Dillon Round (1961/62), the Kennedy Round (1964–67), the Tokyo Round (1973–79) and the Uruguay Round (1986–93). In the course of history, GATT has essentially dealt with tariffs, the major form of trade distortion in the immediate post-war period. All rounds have indeed achieved substantial reductions in tariffs for manufacturing products. NTBs started being addressed under the Tokyo Round, and as a consequence, rules on dispute settlement, antidumping, subsidies, government procurement and intellectual property have been an integral part of the agreement since the 1970s. Rules to protect intellectual property and the opening up of the world market to investment were embodied in the last GATT round. Agriculture was never fully incorporated under the GATT rules until the last round (1986–93), as it was accepted that agriculture was 'different'. The Tokyo Round (1973–79) did contain an agricultural component, which did not nevertheless affect the structure of the Common Agricultural Policy (CAP).

Although an agreement was reached on grain, dairy products and beef, CAP was granted exceptions from the general principles of the agreement and some countries were allowed specific derogations. Some of these derogations and exceptions were incorporated into GATT at the request of the USA (Guyomard *et al.*, 1993). In contrast, the UR has been the most comprehensive and ambitious of all rounds, by placing agriculture at the heart of the negotiations. It could be argued that the incorporation of agriculture under the GATT has permitted a better participation of developing countries in the global trading system, although the treatment of agriculture in GATT negotiations has been excessively dominated by EU/US disputes.

The major institutional change embodied in the Final Act of the UR Agreement concluded in Geneva in December 1993 and signed in Marrakech in April 1994, is the establishment of the WTO, the role of which is to oversee the implementation of all new world trade agreements. The WTO began operation on 1 January 1995 and a new round of agricultural negotiations in the framework of the WTO was initiated in March 2000 in accordance with article 20 of the WTO Agreement on agriculture. The November 2001 Ministerial Conference in Doha decided to launch a new WTO round and to incorporate into it the negotiations on agriculture which had already been mandated in the conclusions of the UR.

Starting with an analysis of agricultural negotiations in the ambit of the UR, this chapter will conclude by highlighting the main issues likely to be dealt with in the current liberalization process. The emphasis of the discussion will be on the EU and the USA as the two main and most powerful negotiators.

8.1 The Uruguay Round of negotiations – the measurement of trade distortions

8.1.1 General background

The UR of multilateral trade negotiations, which commenced in 1986 and was due to be concluded in 1990, has been the most ambitious round ever undertaken. Involving up to 125 countries (108 at the beginning), it covered over 90 per cent of world trade. The UR encompassed 15 areas of negotiations, of which are technical barriers, agriculture, as well as new areas such as financial services and trade-related investment measures. In addition, trade-related intellectual property rights are dealt with under a different council.

The rationale for including agriculture in the UR is explained by the specific world agricultural landscape observable prior the beginning of the UR, that is in the early 1980s. There were three dominant issues at the time. World markets for agricultural commodities were depressed; the US farm sector was in crisis, a situation partly attributable to the overvalued US$, and partly caused by the adverse effects of the CAP;[2] finally, there was a worldwide dissatisfaction with agricultural policies, and their trade distorting effects. Consequently, the objectives of the UR, as stated in the Punta del Este declaration were (CEC, 1994):

(i) to improve market access,
(ii) to ameliorate the competitive environment by increasing the discipline on the use of all direct and indirect subsidies and other measures affecting agricultural trade, and
(iii) to reduce the adverse trade effects of sanitary and phytosanitary regulations.

Although there was agreement on the necessity to include agricultural trade in the round, there were however various degrees of enthusiasm regarding this project. Scheduled to last four years, the UR of negotiations went on for more than six years, essentially because of the EU–US impasse on agricultural trade liberalization. It was clear that from the onset, the US and the EU had radically different objectives. The US, who were joined by other relatively efficient agricultural producers were committed to gaining market access in the EU. The EU, which had accepted the necessity to reform its CAP more substantially, was nevertheless keen to adjust very gradually to a more market-led environment. From the beginning of the UR, it became obvious that the possibility for the US and the EU to reach an agreement on agricultural issues would largely determine the outcome of the GATT round as a whole. This is why a complete understanding of the GATT negotiations compels us to remind briefly the structure of EU–US agricultural trade flows before the beginning of the UR. Focusing on EU–US agricultural trade explains why the USA insisted that agriculture be included in the new round, and why the EU expressed only an extremely limited enthusiasm with regard to this initiative.

The trade balance in agricultural products between the two blocs has historically been in favor of the USA. For example, in 1992, US agricultural exports to the EU reached 6775 million $ whereas US imports from the EU stood at 4434 million $ (CEC, 1994). As was inferred in the previous chapter, the composition of bilateral trade flows between the two blocs is however substantially different. The US competitive advantage is in crops. Grains, animal feed, oilseeds and related products account for nearly half of US exports to the EU. The EU exports to the US are primarily in wine, malt beverages, dairy products, meat, nuts and vegetables. These products account for more than 50 per cent of total EU exports to the USA. The tension between the two countries revived when the 'EU turned to a net exporting position in grains in the early 1980s' (CEC, 1994: 8). During the negotiations, the two major opposing protagonists, the EU and the USA, attracted allies in their respective camp. As a result, two broad diverging views were expressed during the talks: a pro free-trade view and a more cautious approach to trade liberalization. The first line of thinking, which was vehemently defended by the USA, attracted those agricultural exporters that had implemented low levels of protection and support, that were pushing for the reduction and elimination of trade-distorting subsidies, and that could expect tangible terms of trade gains from the new wave of multilateral trade liberalization. Known as the Cairns Group, this group included: Australia, New Zealand, Canada, Argentina, Chile, Colombia, Fiji, Uruguay, Venezuela, Malaysia, Philippines, and at a later stage Hungary. Since the beginning of the talks, the US advocated a 'zero solution', that is a complete liberalization of agricultural trade within a decade (1986–96) (Mahé and Tavéra, 1989), as well as a complete elimination of farm programs by the year 2000 (Guyomard *et al.*, 1993). On the other hand, a more cautious approach to trade liberalization was the view supported by the EU-12 at the time, with support from EFTA countries (Austria, Finland, Iceland, Sweden, Norway), Japan, other importer developing countries, as well as from Canada on issues relating to animal products.

8.1.2 Quantitative protection measures in a comparative perspective

In order for countries to make concrete proposals regarding the nature of trade barriers to dismantle, and in order to allow for a quantitative modeling of the welfare effects of trade liberalization, indicators measuring the extent of distortions had to be devised and commonly accepted. Nominal protection coefficients *(NPC)* measuring the domestic-to-border price ratio (see Chapter 4) involved the calculation of *PSE* and *CSE* indicators (producer and consumer subsidy equivalent respectively). However, comparable *NPC, PSE* and *CSE* estimates for the main traded agricultural products across the major countries of the world were not available when the Uruguay round of talks started (Anderson and Josling, 1993). The first comprehensive collections of such data were commissioned and assembled by both the World Bank for its 1986 World Development Report, and by the OECD in 1987. Since then, the OECD updates its *PSE* series on a regular basis in including new countries in the analysis. A definition of *PSE* provided by the OECD is the following:

$$PSE = (P_d - P_w) \,.\, Q + B$$

where:

PSE is defined as transfers from domestic consumers and taxpayers to producers,

P_d is the domestic price,

P_w is the world price,

Q is the output, and

B are import subsidies, taxation concessions, direct payments linked to the production minus levies on production, as well as cost reducing services such as R&D grants.

A synoptic presentation of *PSEs* across a number of major world agricultural traders and during the 1980s is provided in Table 8.1.

Expressed as a percentage, the *PSE* measure represents that part of receipts accounted for by assistance of various kinds. Therefore, 44 per cent of the producer receipts came from transfers from agricultural policies in 1990. As confirmed by Table 8.1, average protection in the agricultural sectors of the developed countries is virulent, ranging nevertheless from relatively little protection (New Zealand and Australia) to extremely high levels of transfers (Japan). Protection of the then EC agricultural sector is close to the average of the developed world whereas that of the USA is consistently below. In contrast, as documented by Krueger *et al.* (1988) agricultural prices in the developing world, with the exception of rice prices, tend to be below world levels (see below). Rice and dairy products benefited from the highest levels of protection, whereas oilseeds were the least protected products of the group.

The concepts of *PSE* and *CSE* are static in nature. In particular, *PSE* measures imply that what is earned by domestic producers, in excess of what is possible under free trade, is lost by both domestic consumers and taxpayers. Calculated at the level of a domestic economy characterized by a zero-sum game, these indicators entail that trade liberalization would operate a redistribution of wealth from a category of agents to another group of agents. These indicators allow only for a partial and conservative account of the impact of trade liberalization. The resources

Table 8.1 Producer subsidy equivalents (PSEs) for agricultural product – a cross country comparison (1979–90) (in % of receipts)

	1979–86	1987	1988	1989	1990
Australia	12	11	9	10	11
Canada	32	49	42	41	41
EC-10*	37	49	46	48	48
Japan	66	76	74	68	68
New Zealand	25	14	7	5	5
United States	28	41	34	30	30
Average of above countries	37	50	46	44	44
Average by commodity					
Wheat	31	61	46	30	47
Coarse grain	28	56	43	35	39
Rice	74	90	85	82	83
Oilseeds	15	33	27	27	31
Sugar	49	73	62	47	53
Milk	60	70	62	60	68
Beef	41	44	49	44	43
Crops, average	36	62	51	43	49
Livestock, average	37	43	43	40	42

* This excludes Austria, Finland, Portugal, Spain, and Sweden.
Source: Organisation for Economic Cooperation and Development, *Monitoring and Outlook of Agricultural Policies, Markets, and Trade*. Paris, May 1992.

wasted in lobbying efforts by farmers as well as the welfare loss of other nations are, for example, not included in these estimates. However, trade liberalization generates *static* as well as *dynamic* gains. *Static* gains embrace direct consumer and producer effects, whereas *dynamic* gains result from accelerated economic growth rates (see below). Trade liberalization fosters a favorable investment and innovation environment that induces higher growth. More comprehensive calculations of the costs and benefits associated with agricultural policies and including the welfare loss of other nations are displayed in Table 8.2.

The 'gross producer benefits that offset policies of other nations' are all those benefits enjoyed by relatively protectionist nations that distort production and trade conditions in other countries. The latter are essentially the more free-trade inspired nations of the world, as well as the developing world where domestic prices are normally below international levels. Protection in the industrialized nations of the world, such as those represented in Table 8.2 plays as a major deterrent

Table 8.2 Costs and benefits of agricultural support, 1986 (billion of US$, except for last two lines)

Cost/benefit	United States	Canada	EU	Other Western Europe	Japan
Taxpayer and consumer costs	36.3	6.1	46.2	10.6	33.4
Gross producer benefits (GPB)	26.3	3.7	33.3	8.8	22.6
GPB that offset policies of other nations	11.3	2.7	12.5	2.5	1.5
Net producer benefit	15.0	1.0	20.8	6.3	21.1
Gross benefit/costs	72%	61%	69%	83%	68%
Net benefits/costs	41%	16%	43%	59%	63%

Source: Roningen and Dixit, US Department of Agriculture (1989).

for imports from third countries. In particular, the agricultural policies implemented by the EU and by the US in 1986 were responsible for a welfare loss experienced by other nations of nearly US$24 billion. In the early years of the new millennium, agricultural subsidies in developed countries were estimated to be around US$300–350 billion (FAO, 2002). These subsidies were in turn estimated to induce a foregone agricultural income for developing nations roughly equivalent to US$50 billion. Protection in rich countries defeats totally the purpose of agricultural aid policies, for which a meager US$8 million worldwide was allocated towards aid during the same period of time (FAO, 2002). In other words, the cost of support to the agricultural industry in rich countries is substantially higher than the same countries' aid to the developing countries. This more comprehensive framework is taken as the basis for the calculations of the impact of freer trade at the global level.

Before we come to the estimates of the gains arising at the global level from agricultural trade liberalization, we will briefly expose the theoretical foundations on which these estimates are founded. Simple neoclassical economic models are used to show that an expansion of trade through trade liberalization leads to an increase in the level of welfare at the world level. Based on Ricardo's comparative advantage theory (see Chapter 7), these models have been developed since the 1950s and have laid down the foundations of the field of 'regional economic integration'.

8.1.3 Theoretical effects arising from the removal of world trade barriers

Most economic analyses undertaken during the late 1980s and early 1990s concluded that the UR, by liberalizing world trade, would raise overall world welfare. In line with Ricardo's comparative advantage theory presented concisely in Chapter 7, it was widely accepted that trade liberalization would allow a country to shift its resources to areas where they were more efficiently used, releasing them from the production of goods that were produced more cheaply abroad. The impact of the removal of tariffs on the domestic economy, on the economy of associated or partner countries, as well as on the economy of the rest of the world is at the core of the field of 'economic integration', which considers a customs union (CU) as being a limited form of trade liberalization. According to the definition given by GATT, a CU is a geographically confined free trade zone, meeting two basic requirements. First, the participating countries must remove tariffs and other forms of trade restrictions with equivalent effects, among themselves. Second, the member countries define and pursue a Common Trade Policy with the adoption of a Common External Tariff (CET) on imports originating from non-participating countries (GATT, 1952). A CU is an intermediate form of economic integration lying between a Free Trade Area (where no CET exists), and a Common Market, where factors of production (that is labor and capital) are free to move within the integrated area and where various economic policies are developed in common. These various forms of regional economic integration are all geographically bounded, in that they imply internal free trade and external protection. The first consistent theoretical analyses of CU theory appeared simultaneously with the writings of Herbert Giersch (1949/50), Maurice Byé (1950) and Jacob Viner (1950). Feeding upon Ricardo's comparative advantage theory, Viner's CU theory distinguishes between 'trade-creating' and 'trade-diverting' effects of a CU. Using a simple two-country model, Viner (1950) shows that the removal of tariff barriers between two countries leads to overall positive effects (see Chapter 3). The theory also warns against the fact that some agents – firms or groups of consumers – may be adversely affected. Not every one will gain from a trade creating CU. The theory developed in the 1950s places a large emphasis on the static gains arising from freer trade between nations, to the expense of the dynamic gains that result from increased rates of economic growth. Dynamic effects arise because of the improved investment and innovation environment that trade liberalization creates. Moreover, the removal of trade barriers at the world level allows a national agricultural producer of say, beef, to consider the market as being global as opposed to national. A larger market permits in turn the exploitation of economies of scale;[3] it generates higher production volumes at lower per unit costs, thereby creating a more efficient economic environment. This clearly has an impact on the industrial structures of countries and on their production conditions. Again, non-competitive farmers, that is those producing above the minimum efficient scale, will exit the agricultural sector, and the freed resources can be used more profitably in other industries. It is only in the 1970s that the issue of economies of scale became integrated in the CU theory, with the work of Corden (1972). In recent years, the theory has been fine-tuned to incorporate *some* of the elements characterizing a situation of imperfect competition, with the introduction for example of product differentiation in the analysis. Finally, as many of the trade barriers take today the form of NTBs, a modernized theory shows that the gains arising from trade liberalization are even greater when barriers are NTBs, compared with the situation of tariffs (Gowland and James, 1991).

By generalizing the case of the CU between a selected number of countries to world trade liberalization, it follows that the removal of trade barriers globally increases the welfare of all nations. By removing trade barriers, welfare increases at the level of the CU as a whole (case of a CU) or globally (case of multilateral trade liberalization). However, it is important to remember that detrimental effects of trade liberalization will be felt by some categories of economic agents. Therefore trade liberalization, and in particular the UR, will be socially and politically acceptable if and only if the adjustments costs borne by some categories of agents are dealt with in an equitable manner. Theoretically, and assuming that agricultural production in the world is characterized by a perfectly competitive model, the reform of national agricultural policies generated by trade liberalization in the agricultural sector, should, on the whole, lead to a better domestic and international allocation of resources. It was expected that the reordering of agricultural systems would reduce incentives to overuse polluting chemical inputs, and would decrease production on low-yielding fragile land, all changes benefiting the environment (OECD, 1995).

8.1.4 The negotiating position of the EU and the USA[4]

The USA and the Cairns group wanted to achieve a rapid reduction in overall agricultural support and protection. Based on a harmonized measure of agriculture protection developed by the OECD (the famous *PSE*), the USA proposed a 75 per cent cut in farm support, a 75 per cent cut in import protection, and a 90 per cent cut in export support over ten years, from 1991. On the other side of the negotiating table, the December 1990 EU proposals covered two major headings: internal support (intervention) and import protection (variable levies and export refunds). Under the internal support heading, the EU proposed to reduce the aggregate measure of support (*AMS*), another measure of internal support, by 30 per cent from 1986 to 1995/96, claiming however credit for support reductions implemented since 1986. It should be noted that the definition of the *AMS* is less stringent than that of the *PSE*. *AMS* is defined as:

$$AMS = (P_d - P_w) \cdot Q$$

For example, in the cereals and sugar beet sector, the EU offered a 4 per cent annual cut in aggregate support, which over the years 1991–96 would have come to a 18 per cent decrease, on a cumulative declining basis. In the livestock sector (milk, beef, sheep/pig-meat), the offer was an annual 4.7 per cent reduction during five years, leading to a cumulative decline of 20.5 per cent. With regard to import protection, the EU was offering to reduce total support, but it did not wish to have to negotiate on any specific instrument of support, such as for example the much decried export refunds. The EU agreed to convert the variable levy system into a fixed tariff, and to reduce these tariffs by 30 per cent over the five years, that is up to 1996. Consequently, by 1995, tariffs were expected to be £78 per ton for barley, £1600 per ton for butter and £1490 per ton for beef. The EU was careful enough to leave these tariffs at a sufficiently high level so as to keep out normal commercial imports. In 1990, the export refunds had reached a relatively high level; they attained for example at the time 60 per cent of the effective intervention price in the case of beef. The initial position of the EU during the negotiations was a *status quo,* as the system of degressive tariffication newly introduced, as well as the agreed declining intervention prices, was seen to put a natural ceiling on export refunds. Finally, another characteristic of the Commission's proposals was the idea of 're-balancing'. At the time, some products, such as cereals, beef, dairy products, were strongly protected, whereas other products could enter the EU market with little or no import barriers. This was the case for cereal substitutes and oilseeds. The EU offered to re-balance the protective system, that is: to lower import protection on cereals (highly protected) in return for higher import protection on cereal substitutes.

These claims and positions highlight clearly the major fundamental divergences between the two major contracting parties. The gap between

US and EU proposals was far too wide to enable any agreement to come to fruition, and as a result, no agreement was indeed reached in December 1990. Re-balancing was strongly opposed by both the USA and the Cairns group, and the USA expressed a major concern with respect to the EU export subsidies. In the worlds of the EC Commission:

> A complete breakdown was avoided by the offer of the GATT General Director Arthur Dunkel, to prepare a further compromise paper, [...] which by the end of 1991 was accepted as the basis for further negotiations (CEC, 1994: 73).

In the meantime, the EU began serious internal discussions on a substantial reform of the CAP, which, as was seen in Chapter 4, culminated with the June 1992 reform.

8.1.4.1 New iterations (December 1990–July 1992) and the need to reach a compromise

Sent back to its drafting table, the EC Commission proposed a series of transitory arrangements during the course of 1991. After agreement by the Council of Ministers, the Commission defined the agricultural prices for the 1991–92 campaign on the basis of those of the previous campaign (1990–91), by encompassing new price cuts (Baudin, 1991).

This second period in the negotiations was marked by a world economic recession. Although the Gulf war did not have any major implications on production costs in the EU, it did suspend EU exports to countries in the Middle-East, thereby aggravating the problem of stocks. In addition, political and structural change in Eastern Europe, and the ensuing Europe Agreements (Chapter 6) generated a large inflow of beef into the EU in 1990–91, paid at twice the price prevailing in Eastern markets. Consequently, large stocks of beef accumulated in the EU during this period; in January 1991, they reached over 700,000 tons. The situation of oversupply was not confined to the beef sector. It affected also the wheat and sugar sectors, where economic recession translated into sluggish demand. Against this background, the Commission formulated new objectives and guidelines for future policy (CEC, 1991), leading to the June 1992 agreement on the reform of the CAP (OJEC, 1992) (see also Chapter 4). Contained in a global approach, these new proposals reiterated the need to move towards a market-led approach, through a strict management of markets and a reinforced action on prices. Declining prices, and therefore declining incomes for farmers were to be cushioned by an income compensation scheme to EU farmers. Production of arable crops (cereals, oil seeds – that is soybeans, sunflower seeds – and protein crops – peas, field beans) would be curbed further by set aside schemes. These set aside requirements were applied from the 1993/94 marketing year onwards. They concerned 15 per cent of the land used for arable crops. The land set aside was subject to rotation. Here again, the farmer gets compensatory payments in return.

8.2 The 1992 Blair House pre-accord, the 'oilseeds' dispute and the Marrakech agreement

The compensation scheme offered to EU farmers was first strongly opposed by the USA. However, it was subsequently accepted in return for a sizeable cut in export refunds; although the USA had asked for a 24 per cent cut, the agreement was finally settled with a 21 per cent decrease. One of the major contentious points during these negotiations was the oilseeds dispute, which at the time was though to lead to yet another EU–US trade war. It was also an item of negotiations which failed to reunite all EU member states, with for example the French threatening to take an isolationistic position.

Characterized by low self-sufficiency rates (0.70 in 1989 for the EC-12, and still below 1 during the UR negotiations), oilseeds had been subject to special measures in the EU.[5] According to article 2 of the 1982 Council Regulation, prices for peas and field beans (including oilseeds) were in fact increased each month until the early 1990s, amplifying thereby the level of protection for EU oilseeds producers. For example, the monthly increases in the minimum price were fixed at 0.158 ECU per 100 kg for the 1992/93 marketing year. As a result, the agricultural utilized area (AUA) for oilseeds in the EU increased by 38 per cent over the period 1985–89, and by 18 per cent in 1990 only, to 5.263 million ha (CEC, 1992). In 1992, the AUA covered by oilseeds had reached some 5.5 million ha (Le Monde, 1992). Aware of the necessity to comply with the GATT rules and to make appropriate changes in the oilseeds system, the EU reformed the oilseeds regime through the introduction of set aside policies. As a result, producers of oilseeds and of protein crops (peas and beans) were eligible for relatively high rates of compensation. At the time, the price for oilseeds was 2.1 times the cereal prices, and the compensation scheme offered to oilseeds producers still proved to be a favorable treatment. Dissatisfied with the favorable oilseeds regime prevailing in the EU at the time, a special negotiating group was set up in 1988 at the request of the USA to examine the specific problems related to trade in oilseeds.

US dissatisfaction with the EU oilseeds system was fueled by the importance of this commodity as a traditional export source for the USA (Chapter 7), and also by the increasing difficulties encountered by American producers in recent decades. During the 30 years spanning from the Second World War, the USA have largely dominated this market. Charvet (1997) notes how, ironically, US dominance in the world soybeans sector has been made possible by heavy duties and taxes imposed on competing imports since the 1930s. The relatively high level of protection enjoyed over a few decades by American soybeans producers, has enabled the development of the industry. This dominance started being eroded in the 1980s, when world demand became sluggish, and when exports from new competitors (such as Brazil and Argentina) started flooding the world market. As a result, the US world share of soybeans decreased from 83 per cent in 1972/73 to 64 per cent in 1995/96 (Charvet, 1997). Consequently, the powerful American Soybean Association, one of the most influential lobbying groups in the USA, exerted its pressure on the then Reagan administration with the intention of gaining market access in European countries. The US aimed at substantial cuts in the European AUA for these crops. From a European point of view, this had little economic rationale since the self-sufficiency rate for these products was still below 1 at the time.

As seen above, EU internal discussions to reform the CAP, and leading to the 1992 CAP reform, led finally to the Blair House pre-accord on 20 November 1992. This pre-accord laid down a basis for the agricultural aspect of the final UR agreement signed subsequently in Marrakech. First sign of a compromise reached on this very sensitive issue by the two major parties, the November 1992 Blair House Agreement covered the following areas (CEC, 1994):

(i) The AUA for oilseeds should be reduced to 5.128 million ha in the EU (this corresponded to the average AUA over the period 1986–90).
(ii) Subsidized exports would be reduced by 21 per cent, from the average 1986–90.
(iii) Tariffication of all NTBs (see below).
(iv) Reduction of domestic support, that is of guaranteed prices enjoyed by EU farmers.
(v) A 'Minimum accession' principle was laid down: it was accepted that each member country would have to open up its market further to foreign produce. The increase of imports would be of 3 per cent of total domestic production at first, and 5 per cent at a later stage.

France, a relatively large producer of oilseeds, was amongst the EU countries the most hit by this pre-accord. France was a major player in the 'oilseeds' dispute, and it was probably the EU country that made its voice heard the most vehemently during the negotiations in looking for major compensations.[6] The French finally accepted the oilseeds deal in return for a retention of a protectionist stance in their audio–visual industry. It was estimated at the time that the Blair House pre-accord would have resulted in a 5 per cent decrease in EU market shares for milk, and a 50 per cent decrease in exports of pork. Exports of wine would have decreased by 450,000 hl, compared with 2.2 million hl produced in 1992, corresponding to a 20 per cent decrease. All these are important exports for French agriculture.

8.2.1 Concluding the UR – the 1994 Marrakech agreement

As can be seen, many iterations in the talks led to the final agreement being concluded in Geneva in December 1993 and signed in Marrakech in April 1994, and this ended the cycle of the UR. The agreement was implemented at different dates throughout 1995, with the entry into force of the WTO, after domestic ratification in all the countries concerned was completed, so that the commitments undertaken could be given legal standing. The Marrakech agreement was acclaimed widely among economic and political circles. According to the OECD (1995: 161), the agreement on agriculture 'contains an innovative attempt to discipline those domestic policy measures deemed to be the most distorting in terms of production and trade'. It is possible to distinguish the external and internal components of the UR agreement on agriculture (Table 8.3).

8.2.1.1 External or trade impact

The most immediate policy changes were obviously felt at the level of trade measures. It was agreed that all tariffs discriminating against agricultural imports would be 'bound', against future increases, and that all existing NTBs to trade (including quantitative restrictions, VERs and import bans) would be converted into tariffs, a process known as 'tariffication'. Under GATT/WTO, the contracting parties were requested to 'bind' their mutually negotiated tariff concessions, implying that these 'bound' tariffs cannot be amended without new multilateral negotiations (Senti and Conlan, 1998). The 1986–88 period was the reference period for the calculation of the reduction in protection after the UR. As these years were marked by specific world market conditions, it was felt that 'tariffication' could actually generate high tariff protection. As a result, developed countries agreed to offer minimum market access opportunities

Table 8.3 The main components of the 1993 UR agreement on agriculture

Level of impact	Main components
External impact	Tariffs are bound; process of 'tariffication' and introduction of tariff rate quotas
	Average tariff cuts of 36% by developed countries; minimum access opportunities
	Subsidized exports in developed countries must be reduced by 21% (compared with a base of 1986–90), and expenditure on export subsidies by 36%. Less stringent cuts apply to developing countries
Internal impact	Trade distorting domestic policies are reduced by 20%

to a number of politically sensitive commodities. This involved the introduction of a compromise policy instrument that took the form of tariff quotas at reduced rates for a share of the domestic consumption. This is why tariffication involved in many cases the establishment of tariff rate quotas (TRQs), with specified access levels being provided at lower duties (OECD, 1999). Developed countries offered tariff reduction commitments of 36 per cent, compared with the base of 1986–88. A gradual liberalization process, through cuts in bound tariff has therefore been under way in agriculture since 1 January 1995, date of entry into force of the WTO. For industrialized countries, the final tariff rates have been in effect since January 2000, whereas developing countries have been accorded a period of ten years (that is until January 2004).

The variation of post UR tariffs in agriculture is still large, with very high mean tariffs (that is more than 100 per cent) in countries such as India, Norway and Tunisia. This contrasts with extremely low mean tariffs for agriculture in Australia (3 per cent), Canada (5 per cent), the USA (6 per cent) and New Zealand (9 per cent). By comparison, the EU mean tariff on agricultural products was 20 per cent in the late 1990s (OECD, 1999).

With regard to export measures (that is export refunds and subsidies), the agreement provided for a reduction in expenditure on export subsidies of 36 per cent, as well as a reduction in the quantity of subsidized exports by 21 per cent during the six-year implementation period. The figures for developing countries are 24 and 14 per cent respectively. These measures applied to direct subsidies as well as to other payments such as marketing cost subsidies and specific transport subsidies. For the increased access provisions, countries needed to find an allocation mechanism consistent with the most favored nation principle. Note that some of the minimum access provisions were already accounted for under various bilateral or preferential agreements. Finally, in terms of sanitary and phytosanitary measures, the agreement sets out clearer and more detailed rights and obligations for food safety and animal and plant health. However, the agreement warns that these rights and obligations should not be used as an arbitrary means of trade discrimination. In addition, the agreement calls for a harmonization of standards based on international best practice. The Codex Alimentarius Commission, the International Office of Epizootics are for example mentioned in the agreement.

8.2.1.2 Internal or domestic policy impact

The major aspect under this heading is the reduction in domestic support. Assistance, which is more than minimally production and trade distorting, and as measured by *AMS*, is subject to a 20 per cent reduction commitment over the six-year implementation period from 1995 to 2000 for developed countries. A 13 to 14 per cent reduction applies to developing countries. Within the terms of the agreement, it is however possible for countries to compensate fully for reductions in price support through direct payments. These measures were thus to switch the burden of assistance to taxpayers, away from consumers.

A *Decision on Measures concerning the Possible Negative Effects of the Reform Program on Least-Developed and Net Food-Importing Developing Countries* was also included in the agreement. These mechanisms are aimed at ensuring that the implementation of the measures above does not adversely affect the availability of food aid, so that assistance can still be provided to the least-developed and net food-importing countries. In addition, the 'Green' and 'Blue Boxes' contain a range of policies exempted from the commitment to reduce domestic support. The rationale for exempting these policies stems from the fact that they are deemed to be non- or minimally production and trade distorting.

8.3 The CAP–GATT/WTO link

During the first part of the UR, the official position in Brussels was, through its Commissioner for Agriculture Mac Sharry, to revoke any causality between the GATT talks and the reform of the CAP. Since agriculture had been a major obstacle to the conclusion of the UR in 1990, it is obvious that without a substantial redrafting of the December 1990 EU proposals, so as to make the CAP more compatible with international (mostly US) obligations, the attempt to reach a final agreement would have failed. A proof is that, shortly after the drafting on 1 February

1991 of new proposals by the Commission for the future of the CAP (CEC, 1991), the GATT General Director announced (on 19 February) a resumption in the multilateral talks. In addition, the USA had in fact insisted that failure to reach an agreement in the Agriculture negotiating group would prevent the successful termination of the entire UR, jeopardizing progress achieved in 14 other negotiating groups. It is undeniable that the developments within the framework of the GATT during the 1980s, created additional pressures to reform substantially the CAP. The content of the 1992 reform is evidence that multilateral trade negotiations exerted an influence on Brussels.

It should however be noted that although there is a link between the GATT negotiations and the reform of the CAP, in that the former accelerated the pace of CAP reform, this link has probably sometimes been exaggerated. Indeed, the reform of the CAP would have taken place notwithstanding the external elements represented by the espousal by GATT of *laissez-faire* economic principles surging again at the global level. As has been described earlier (see Chapter 4), the 1992 CAP Reform represents the culminating point and also a more radical step of a long-lasting process that had started in the 1970s. That the GATT multilateral negotiations increased the speed of the CAP reform is therefore unquestionable.[7]

8.4 Estimating the effects of trade liberalization

8.4.1 Ex-ante analyses

Quantifying the complex trade interactions between more than 100 countries is a much more arduous task than that represented by the estimation of trade gains arising in the case of a simplified two-country CU model, *à la* Viner. However, in using computable general equilibrium analyses, *ex ante* estimates of gains resulting from the UR have been undertaken by various research bodies and international organizations. In spite of the diverse assumptions[8] used and of the many different results obtained, the following broad common picture emerges. Most net exporters of temperate-zone agricultural products such as Australia, New Zealand and in some cases the USA, were to benefit from easier market access. Most net importers, such as Japan and some of the EU countries, would be able to gain through the replacement of high-cost domestic production by lower-cost imports, but would lose from adverse terms of trade effects. Developing countries should in general enjoy higher agricultural exports and should reduce their imports of agricultural products.

In particular, a joint-study by the World Bank and the OECD (1993) estimates that partial trade liberalization in both the manufacturing and agricultural sectors could add some US$213 billion to global income on an annual basis and beginning in year 2002.[9] Of these overall income gains, the 24 OECD industrial countries would reap 63 per cent, or $135 billion. In the non-OECD area, losses would be around US$7 billion; these losses would represent however a small fraction of the gains accruing to these countries estimated at US$85 billion. In particular, the developing countries as a whole would gain US$70 billion. Gains and losses arising from agricultural trade liberalization alone would amount to US$120 billion and US$2 billion respectively in the OECD area. Clearly, most of the gains derived from trade liberalization would originate from the removal of trade obstacles in the agricultural sector. The losses are only a fraction of the expected gains. It is worth analyzing the respective position of countries, starting with the developing world.

8.4.1.1 Effects on the developing world

First of all, it should be noted that the number of developing nations accepting the obligations of the GATT/WTO grew from 61 in 1980 to 110 in 1999 (Killick, 2001). One such *ex-ante* study by Balassa (1988) predicts that the developing countries would be the major beneficiaries of global trade liberalization, as they have been the victims of depressed world prices, caused by plethoric supply from highly protected developed countries' agricultural systems. In an early study quantifying the gains to the developing countries arising from global trade liberalization, Valdès and Zietz (1980) estimated a net gain in economic welfare to the developing countries of US$0.9 billion, over the years 1975–77. The authors' study is based on the assumption that average OECD tariff rates would decrease by 50 per cent. According to their results, the developing countries would enjoy considerable increases in world market shares. They foresaw large gains for commodities growing in tropical areas, such as sugar, tobacco, coffee, cocoa, oilseeds and oils. The gains would however be substantial also for other temperate-zone foods, of which the industrial countries are important exporters; beef, lamb, pork and poultry would all benefit from substantial increases in world market shares.

The most obvious measure that distorts trade conditions and production in the developing world is the tariff structure of processed versus non-processed commodities. Looking at the whole agricultural *filière*, Yeats (1981) estimated that average tariffs facing developing countries exports in the industrial countries were as follows in the 1970s:

8.9 per cent on fresh vegetables	against	12.4 per cent on prepared vegetables,
4.8 per cent on fresh fruits	against	16.5 per cent on prepared fruits,
2.7 per cent on oilseed	against	8.1 per cent on vegetable oils,
6.8 per cent on coffee beans	against	9.4 per cent on processed coffee, and
2.6 per cent on cocoa beans	against	4.3 per cent on processed cocoa.

Clearly, the fact that relatively high tariffs have been imposed on imports of processed food from developing countries has mitigated against the development of a food industry in these countries, preventing these countries from diversifying their exports.

However, according to Killick (2001), welfare gains for the developing world should not be exaggerated. For the developing world as a whole, welfare effects were nevertheless though to be small, representing typically less than 1 percentage point of GDP.

Moreover, the optimistic scenarios above had to be toned down by the extreme diverse situations characterizing the developing world. It is clear that not all developing nations gain from agricultural trade liberalization, and that the probability to gain depends very much on the net trade position. Trade liberalization leads theoretically to substantial gains for net food exporting developing nations. As more than half of the developing nations were net food importers in the 1990s, most of these nations were therefore bound to lose from the GATT/WTO rounds (Killick, 2001). This has been particularly the case for African and Middle East countries, a situation in sharp contrast with that of Latin America. Most of Latin American countries may be better off, although Brazil is in a less favorable situation as the country has significantly protected its cereals sector, and has taxed heavily tropical crops and livestock. Already in their 1993 joint report, the World Bank and the OECD had warned that most of sub-Saharan Africa, Indonesia and the Mediterranean countries would not perform well. In particular, Indonesia would see its export prices for rice, coffee and cocoa decline whereas the price the country pays for imports of wheat, meat and dairy products would climb (World Bank and OECD, 1993).

8.4.1.2 Effects on the USA and on Europe

Most studies conducted on the case of the USA concluded with the beneficial impact of trade liberalization on the US economy.[10] Elwell and Reifman (1993) provide a review of *ex-ante* studies on the impact of trade liberalization for the USA. For example, in assuming a one-third cut in tariffs and NTBs, as well as an accrual of gains over a 20-year period, the United States Trade Representatives estimated that the US economy would enjoy static efficiency gains of about US$130 billion per year by the end of the 20 years (USTR, 1990). For the agricultural sector proper, potential sizeable gains were foreseen in the area of grain, where higher prices were projected. These gains would more than compensate for the losses in other sub-sectors.

Quantifying the effects of trade liberalization on the EU is an arduous task, as it is indeed not easy to differentiate between CAP reform effects and GATT effects. This is why most *ex-ante* analyses would have tended either to take the two issues together or to focus on the impact of the CAP reform on European agriculture (see for example Sheehy, 1995/96 and Guyomard *et al.*, 1993).

8.4.2 Ex-post analyses

As was also the case for the formation of the European Common Market in the late 1950s, *ex-post* analyses on the observed impact of trade liberalization do not seem to be as numerous as the *ex-ante* analyses. In our specific instance, the paucity of *ex-post* studies on the UR can only be partly explained by the ongoing liberalization process, particularly in the case of developing countries. Actually, economists have already started being preoccupied with the potential effects of further trade liberalization in the agricultural sector. One such study is the recent simulation by Kennedy and Atici (1998) of complete agricultural trade liberalization between the US and the EU with respect to the agricultural sector. The authors use a static and partial equilibrium model, highlighting the case of the EU, the US, and a more politically passive rest of the world. Their results show how total free trade in the agricultural sector affects domestic and world prices, production, consumption, self-sufficiency and welfare. The *ex-post* immediate impact of agricultural trade liberalization can however be gauged at various levels. For example, it is worth noting the possible impact the UR may have had on the level of prices, on farmers' incomes, on export market shares, and on the terms of trade within a number of regions in the world, such as the EU. The case of the developing world, which has not received adequate attention, is also explored under this section.

8.4.2.1 The impact of the UR on EU agriculture

Being a core element of the 1992 CAP Reform, the decrease in guaranteed prices is what best characterizes post UR agriculture in the EU. This is the most notable achievement of the CAP reform/UR for EU countries. In the case of cereals, the intervention price decreased from ECU 155 per ton before the reform, to ECU 100 per ton in 1995/96.[11] The intervention price for beef fell by 15 per cent from ECU 326 per 100 kg in 1993, to ECU 292 in 1995. Reflecting the cuts in intervention prices, nominal producer prices for total agricultural production fell substantially in the 1990s, in all EU countries, with the exception of Greece, Spain and Italy. Indeed, the picture tends to vary substantially across EU member states. In Portugal, producer prices have been almost constant in nominal terms, whereas in Ireland and in The Netherlands, only a slight increase was registered over the period (EUROSTAT, 2002). The levels of stocks for the main intervention commodities have tended to decrease during the late 1990s, leading to important savings made on the EC budget. This is particularly true for butter over the whole period, whereas intervention stocks for cereals fell sharply in 2000 (CEC, 2002). In short, the re-ordering of the EU agricultural sector has been successfully implemented (see also Chapter 4).

Notwithstanding the decrease in producer prices since the implementation of the UR agreement, the predictions that EU consumers would benefit from these reductions did not seem to materialize. The data contained in Table 8.4 show that consumer prices for foodstuff and beverages have actually increased in the EU as a whole since the mid-1990s, a phenomenon in sharp contrast with the evolution of producer prices. The increases have been particularly significant in 1996 and 1998. The taking into account of upward trends in excise duties on alcohols in some countries over the period (such as in Ireland and in the UK) does not provide an adequate explanation for these increases. Excise duties have tended to decline in other EU countries over the period, offsetting the increase referred to above. Food price increases have been particularly significant in Greece, Italy and Portugal, with average yearly price increases of 4.5, 2.6 and 2.1 per cent respectively. As was noted in Chapter 4, Finland, Sweden, and to a lesser extent Austria escape from

Table 8.4 Annual percentage change of consumer prices for foodstuff and beverages in the EU (1995–2000)

	1995	**1996**	**1997**	**1998**	**1999**	**2000**
Producer prices	4.3	0.4	−1.4	−2.5	−3.4	0.9
Consumer prices	–	1.9	0.8	1.4	0.3	0.7

Source: CEC (2002).

this rule, given that membership to the EU and participation in the CAP, meant actually a substantial decline in producer prices in the second half of the 1990s. These declines were so substantial (nearly 25 per cent in both Austria and Finland) that transmission effects down to the final consumer took place eventually.

Undoubtedly and logically, the widening gap between producer and consumer prices in the EU has benefited the intermediary organizations such as the food processing industry, but above all, the distribution sector. Both the food processing industry and the distribution industries in the EU are characterized by imperfect competition. The completion of the internal market in the EU (the so-called '1992 program') led to increasing concentration ratios in these two industries. This shows that agricultural trade liberalization alone does not guarantee lower prices for consumers.

8.4.2.2 The wider context: impact on other (developing) economies

The model simulations performed estimated that the developing countries' market access to developed countries' markets would be improved, thanks to lower tariffs and to the removal of quantitative restrictions. As we have seen in the previous chapter, the developing countries produce a large range of tropical and traditional products (such as meat and dairy commodities), and many of these countries have a comparative advantage in the production of agricultural and food products, compared with developed countries. This implies that, for these countries, trade liberalization matters to a great extent. The export performance of developing countries since the mid-1990s has been very uneven, with Latin America, the Caribbean and Asian countries expanding their exports of agricultural and food products. The fears expressed in the ex-ante analyses have materialized, for Africa lost market shares in all agricultural and food commodities since that time (Henson and Loader, 2001). Besides, although, as noted by Matthews (2001), developing countries in general have been able to increase their world market shares of agricultural products in the 1990s, it is not clear whether this is explained by the decline in protection rather than by the relative decline of EU exports on world markets, a decline explained by curbs in EU production. Although the UR agreement did not lead to higher world food prices in general, as was feared by the developing countries, the evidence showing that the agreement has generated positive effects for these countries is skimpy. For example, the case of Mexico and of other Central and South American countries shows a switch over from food security to export-oriented farm policies. In Honduras, trade liberalization, combined with the structural adjustment policies of the 1980s, two inter-twinned elements, has resulted in higher interest rates and higher costs of credit for farmers (Johnson, 1997).

The general consensus that seems to emerge in the economics literature is that, although the developing countries have acquired a more active role in multilateral negotiations over the years, they are still nevertheless the eternal losers, given their economic and political weakness. For example, producing almost everything, the developing countries do not have a dominant position on the world market, except in the case of coffee, where they account for four-fifths of total exports. Consequently, their bargaining position in the ambit of the WTO is still weak.

Moreover, although tariffs have tended to decline since the mid-1990s, it is worth asking whether tariff barriers have not been replaced by other forms of subtle trade impediments. According to Clark (1998), the fall of tariffs has been paralleled with a greater reliance on the use of other trade barriers by industrial nations. Basing his analysis on US protectionism, Clark (1998) finds that the USA use a wide range of NTBs to restrict agricultural and other imports from developing nations. In particular, and according to the same author, US tariff-rate quotas (TRQs) have hampered sugar and other agricultural imports into the USA from Thailand, the Philippines, Guatemala and the Dominican Republic. The TRQs have indeed been criticized for lacking transparency, like quotas, and for not allowing greater market access, particularly in the case of the developing countries, who have complained about the lack of liberalization following the UR (Abbott, 2002). The text of the UR agreement on agriculture related to the notion of 'minimum access commitment' is loose, and it is therefore subject to flexible interpretations. As a result, the extent of trade liberalization achieved so far in agriculture has been extremely limited, and TRQs in particular have been a 'failed market access instrument' (Abbott, 2002: 109).

Already shortly after the entry into force of the UR agreement on agriculture, authors such as Mahé (1997) had warned against the emergence of a new type of protectionism, one which is based on health, environmental and/or ethical grounds. Henson and Loader (2001) note that the number of notifications of technical barriers, such as food safety regulations and standards, to the GATT/WTO has increased dramatically since the early 1980s. Basing their analysis on the case of Sanitary and Phytosanitary Requirements (SPS) and of market access into the EU, the authors argue that SPS act as an efficient form of trade barriers by prohibiting imports. The survey methodology used in their study allows the authors to conclude that SPS measures and other technical requirements (such as labeling) have been the most important impediments to exports to the EU in the late 1998/99.[12] NTBs are an extremely important issue for developing countries, given their comparative advantage in the agricultural sector, and a lack of technical capability enabling these countries to comply with technical requirements. In particular, food hygiene requirements in the EU are increasingly stringent and complex. GMOs are inevitably a new source of trade barriers, which is nevertheless likely to take prominence in EU–US negotiations within the new WTO round, an issue explored in the next section.

8.5 The future – the November 2001 Doha agenda and beyond

The work of the WTO committee on agriculture, aimed at preparing a new round of multilateral trade negotiations in agriculture was completed in September 1999. The November 2001 Doha Ministerial Conference, which succeeded the widely contested and failed attempt at Seattle two years prior that date, adopted the Declaration launching a new WTO round. The decision of 1 February 2002 in Geneva established the negotiating structure for the new round of WTO trade negotiations. In accordance with Article 20 of the WTO agreement on agriculture, the new round of trade and agricultural negotiations in the framework of the WTO should lead to modalities for further commitments in the course of 2003, and to the conclusion and entry into force of the new commitments in January 2005.[13]

There is today a wide dispersion of opinions between the pro free traders on the one hand, and the skeptics that remain yet to be convinced by the positive effects of trade liberalization, on the other. The USA and many other western governments, still privilege in their overall economic thinking the following sequence of events: further trade liberalization should promote export-led growth enjoyed by the developing world, whose global market shares for the products in which they have comparative advantages, were inhibited by highly protected western agricultural markets. This should finally lead to rising economic growth rates and to rising incomes.

The non-governmental organizations (NGOs), which have become so prominent in recent years, and the farm lobbying groups find it difficult to accept the above argument. They consider that regular food supplies to every individual in every nation of the world is a fundamental human right, and that food security should be paramount. According to this alternative view, and invoking the 'market failure argument', the market is simply seen as being unable to provide an optimal allocation of food supplies. Since much of the transformation of primary food inputs into final food products is in the hands of large food multinationals, interested only in short-term profit maximization, governments have to intervene so as to help in creating competitive advantages, particularly in the developing world. The position of the FAO lies between these two extremes, and the EU still espouses a *laissez faire* attitude tinted however with a touch of philanthropy. Aware of the importance of food and agriculture in the developing world, the EU calls for a special and differential treatment for developing countries. Prepared to go further than Lome, the EU insists that the new deliberations be focused more on the needs of developing countries; this led to its defining an 'everything but arms' (EBA) approach. The EU EBA proposal would extend tariff- and quota-free access to all exports from the poorest 48 developing countries. This would concern the remaining products that are still currently excluded from the current EU's preferential trade regime with developing countries. Page and Hewitt (2002) warn that the EBA initiative could sometimes actually divert trade from poorer countries; the authors see this initiative as merely a political move, rather than as a tool of a well thought development policy. Not surprisingly, market access – for its own exporters, continued the reduction of agricultural support, and non-trade measures are listed as the major concerns of the EU in the new talks. The Community will definitely seek to obtain improvements in opportunities for its own exporters. Based on the Agenda 2000 package, it reiterates its commitment to reduce support, as it wishes to establish a fair and market-oriented agricultural trading system. Non-trade concerns such as food safety, policies to protect the environment and to preserve the countryside, as well as animal welfare are also important issues. The EU wishes to reassure its consumers by confirming that the WTO will not be used to force onto the market products the safety of which would be reason for concern. It should be noted that issues of concern to agriculture can arise under other headings such as SPS agreements, technical barriers to trade and also trade-related aspects on Intellectual Property Rights (TRIPs). Article 5 of the TRIPs Agreement reached under the UR deals with trade marks and brands. In particular, through its emphasis on intellectual property protection, the UR has strengthened the private character of biotechnologies, reinforcing the private nature of business and research in the area. Sheldon (2002) foresees an important EU–US trade dispute arising on GMOs. The author highlights the extreme diverse approaches taken by both the USA and the EU in terms of GMOs. The US approach requires no general requirement for labeling of GM foods. The basic argument underlying this approach is that the zero tolerance principle (for potentially hazardous ingredients in food) cannot be taken seriously, as it would otherwise result in very few food products being marketed. On the other hand, the EU approach is extremely cautious. A product cannot be marketed unless it has undergone an environmental and health risk assessment procedure, and unless it has proper labeling, enabling its traceability (see Chapter 5). Moreover, the EU approach assumes that even if the risk assessment has given rise to positive results, these cannot be treated as conclusive, and therefore, the period of consent cannot exceed ten years. As noted by Sheldon (2002), exporting nations to the EU will complain about the excessively strict EU regulations, whereas the EU is likely to argue that the existing scientific knowledge about GMOs is insufficient to allow free trade to occur.

Conclusions: contentious issues in the new round of negotiations

The main focus of the various GATT rounds in the past has been the decrease in protectionism, albeit more in the manufacturing sector than in the relatively highly protected agricultural sector. Decreasing levels of protection have been visible through declining tariff rates. The UR of trade liberalization was the most ambitious of all GATT rounds, by placing agriculture at the heart of the negotiations. The UR was finally concluded in Geneva in December 1993, after an impasse in 1990 and several iterations. The agreement embodies a comprehensive package of individual member country commitments including tariffication of NTBs, tariff concessions and bindings, reduction in domestic support and export subsidies to be implemented over a six-year period (to 2000), or over a ten-year period to 2004 for developing countries (OECD, 1995). Moreover, the bilateral discussions between the EU and the US in Blair House in December 1992 allowed for the exemption from the agreement of the EU income compensation payments and of the US deficiency payments resulting from the 1985 Farm Bill.

The agreement on agriculture has been in force since 1 January 1995. The most visible impact on EU agriculture has been the decrease in the intervention price for most regulated agricultural commodities, despite the fact that it is extremely difficult to isolate trade liberalization effects from CAP reforms effects in the case of the EU. However, the decrease in producer prices has not led, as expected, to a reduction in consumer food prices in the EU. Consumer prices for foodstuff and beverages continued on their ascendant trend since the mid-1990s. The increases have been particularly significant in countries such as Greece, Italy and Portugal. Moreover, although developing countries were thought to be among the major winners of the round, their extreme diversity led to mitigated results. Agricultural trade liberalization led to uneven gains for the developing world, with net food importing countries, such as African and Middle East countries losing out, a situation in sharp contrast with that of Latin American countries.

Following conclusions of the UR on 15 December 1993, the 1994 Marrakesh Agreement established the WTO with its strengthened Dispute Settlement procedure. The failure to start a new round of multilateral trade negotiations in Seattle in 1999, was followed by the Ministerial Conference in Doha (Qatar) in November 2001. This launched the new WTO round by incorporating into it the negotiations on agriculture, which had already been mandated in the conclusions of the UR. According to the negotiating timetable adopted in Doha, negotiations are planned to conclude in December 2004. The EU negotiating position is based on its Agenda 2000 package, and it highlights three main issues. First, that non-trade aspects of agriculture, such as food safety and quality, be addressed. Second, the EU seeks the improvement of market access for its own food exporters, while being committed to a continued reduction of agricultural support. Third, conscious of the past failure to serve the interests of poorer countries, the EU is adamant that more space be given to the needs of the developing world in the new round of trade negotiations. The EU sees the need for special and differential treatment for developing countries, given the great importance of food and agriculture in these countries. The EU Commissioner Fischler has criticized the USA for the lack of concern shown in relation to the interests of developing countries.[14] The developing countries and the countries experimenting a transition from central planning to a market led economy now represent 80 per cent of WTO members. Willing to allow these countries to benefit from the expansion of world agricultural trade, the EU has adopted an 'Everything-but-Arms initiative', allowing duty- and quota-free access to all products from nearly 50 least developed countries.

In the light of tariffication and of falling tariffs, an important and pertinent question is to ascertain whether tariffs have not been replaced by other forms of trade impediments. Recent evidence has shown that the fall of tariffs has tended to be paralleled with a greater reliance on the use of NTBs by industrial nations. Biotechnology, GMOs, property rights, environmental and ethical issues, are all issues likely to dominate the new round of agricultural trade negotiations.

Key terms and concepts

Multilateral trade negotiations
Agricultural trade liberalization
Welfare effects arising from trade liberalization
Producer subsidy equivalent
Non-tariff barriers
Tarrification
Tariff rate quotas (TRQs)
Intellectual property rights
'Everything but Arms' (EbA) approach

Notes

1 These were: Australia, Belgium, Brazil, Burma (Myanmar), Canada, Ceylon (Sri Lanka), Chile, China, Cuba, Czechoslovakia, France, India, Lebanon, Luxembourg, The Netherlands, New Zealand, Norway, Pakistan, Rhodesia, Syria, Union of South Africa, United Kingdom and USA. Note that after denouncing the GATT, the People's Republic of China was able to enjoy observer status since 1984 (Senti and Conlan, 1998). The country became formally a member of the WTO in November 2001, ending a 15-year negotiation process. Are members of the organization those countries who pay an entrance fee. Functioning is based on consensus and not on majority.
2 In the second half of the 1980s, the depreciation of the dollar helped reduce the large US trade deficit by boosting exports. An economic recovery during this period lessened the need to protect weak and exposed sectors, such as the agricultural sector.
3 Economies of scale are defined as a per unit average cost decrease, as output and the scale of operation expand.
4 All statistical information in this section is from CEC (1994).
5 See Council regulation, No. 1431/82 of 18 May 1982, and also Council Regulation No. 1751/92.
6 For more on the internal divisions in the EC Commission during the talks, see Ackrill (2000).
7 For more on the CAP–GATT link, see Swinbank (1999).
8 Various assumptions have been made on the expected variation of exchange rates, on rates of technological change (assumed to be constant in many cases), and on the magnitude of price cuts and on other policy changes.

9 In 1992 dollars. The model does not take into account trade barriers in the services sector as well as NTBs. For the agricultural sector, the model assumes a 30 per cent cut in agricultural tariffs, export subsidies and input subsidies. The report stresses the fact that full agricultural trade liberalization, meaning the elimination of all forms of intervention, ranging from tariffs to subsidies could add US$ 430 billion to world annual income.
10 An exception to this trend is the study by the Economic Strategy Institute (ESI, 1992) concluding that the Uruguay Round would increase the US trade deficit of US$ 37.7 to US$ 62.4 billion, and that such a deficit would result in a decrease of US GDP.
11 It should be noted that the prices for cereals on world markets, and in particular for wheat, shot up during the mid-1990s (1994–96). In the EU, the 2000–01 prices were down considerably, compared with these record levels.
12 The survey involved government personnel, *au fait* with SPS and WTO issues, in 10 countries namely: India, Egypt, Guatemala, Vietnam, Cameroon, Ethiopia, Ghana, Gambia, Kenya and Zimbabwe.
13 Until the 5th Ministerial in 2003, the Doha Development Agenda will have negotiating bodies for Agriculture, Services, Industrial Tariffs, Trade Rules, Intellectual Property and Environment.
14 See //europa.eu.int/comm/trade/goods/agri/pr260702_en.htm.

Bibliography

Abbott, P.C. (2002) 'Tariff-rate quotas: Failed market access instruments?', *European Review of Agricultural Economics,* 29: 1, 1st quarter, 109–30.

Ackrill, R. (2000) *The Common Agricultural Policy* (Sheffield: Sheffield Academic Press).

Anderson, K. and Josling, T. (1993) *The Challenges to Economists of Multilateral Trade Negotiations on Agricultural Protection* (Food Research Institute Studies, Vol. XXII, No. 3, 275–304).

Baudin, P. (1991) 'La fixation des prix agricoles pour 1991/1992', *Revue du Marché Commun et de l'Union Européenne,* No. 349, Juillet/Aout, 535–51.

Balassa (1988) 'Agricultural policies and international resource allocation', *European Review of Agricultural Economics,* 15: 2/3, 159–71.

Byé, M. (1950) 'Unions douanières et données nationales', *Economie Appliquée,* Janvier-Mars, 121–58.

CEC (1991) *The Development and Future of the Common Agricultural Policy – Proposals of the Commission,* COM (91) 100 final, 1 February, Brussels.

CEC (1992) *Volet agricole de la négociation Gatt et réforme de la PAC* (Commission Communication CEC (92) 2267 final).

CEC (1994) *The Economics of the Common Agricultural Policy,* European Economy No. 5 (Brussels/Luxembourg).

CEC (2001, 2002) *The Agricultural Situation in the European Union,* 1999 and 2000 Reports (Brussels and Luxemburg).

Charvet, J.P. (1997) *La France Agricole – Dans son Environnement Européen et Mondial* (Editions Liris: Paris).

Clark, D.P. (1998) 'Are poorer developing countries the targets of US protectionist actions?', *Economic Development and Cultural Change,* 47: 1, 193–207.

Corden, W.M. (1972) 'Economies of scale and customs union theory', *Journal of Political Economy,* Vol. 80, 465–75.
Economic Strategy Institute (1992) *The Uruguay Round and the United States: A Critical Analysis*. ESI, Washington DC.
Elwell, C. and Reifman, A. (1993) *The Uruguay Round – A Macroeconomic Assessment,* June 1, Congressional Research Service Report for Congress, Washington.
EUROSTAT (2002) *Yearbook 2002 – The Statistical Guide to Europe* (Brussels and Luxembourg).
FAO (2002) *World Food Summit – Five Years Later.* Food Agricultural Organization: Rome, 10–13 June.
GATT (1952) *Basic Instruments and Selected Documents.* Vol. I, Part III, Article XXIV, Section 8(a), Geneva.
Giersch, H. (1949/50) 'Economic union between nations and the location of industries', *Review of Economic Studies,* No. 2, 87–97.
Gowland, D. and James, S. (eds) (1991) *Economic Policy after 1992* (Aldershot: Dartmouth Publishing Company).
Guyomard, H., Mahé, L.P., Munk, K. and Roe, T. (1993) *Agriculture in the Uruguay Round – ambitions and realities.* CEC Economic Papers, No. 101, March, Brussels.
Henson, S. and Loader, R. (2001) 'Barriers to agricultural exports from developing countries: the role of sanitary and phytosanitary requirements', *World Development,* 29: 1, 85–102.
Johnson, H. (1997) 'Food Insecurity as a Sustainability Issue: lessons from Honduran Maize Farming', in J.P. de Groot and R. Ruben (eds) *Sustainable Agriculture in Central America* (New York: MacMillan).
Killick, T. (2001) 'Globalization and the rural poor', *Development Policy Review,* 19: 2, 155–80.
Kennedy, P.L. and Atici, C. (1998) 'A sectoral analysis of agricultural trade liberalization', *Journal of Agricultural and Applied Economics,* 30: 2, December.
Krueger Anne, O., Schiff, M. and Valdés, A. (1988) 'Agricultural incentives in developing countries: measuring the effect of sectoral and economywide policies', *World Bank Economic Review,* Vol. 2, No. 3, 255–71.
Le Monde (1992) Wednesday 25 November 1992.
Mahé, L.P. and Tavéra (1989) 'Bilateral harmonization of EC and US agricultural policies', *European Review of Agricultural Economics,* 15, 327–48.
Mahé, L.P. (1997) 'Environment and quality standards in the WTO: New protectionism in agricultural trade? A European perspective', *European Review of Agricultural Economics,* 29: 3/4, 3rd and 4th Quarter, 480–503.
Matthews, A. (2001) 'Developing countries' position in WTO agricultural trade negotiations', *Development Policy Review,* 20: 1, 75–90.
OECD (1995) *Agricultural Policies, Markets and Trade in OECD Countries – Monitoring and Outlook.* Paris.
OECD (1999) *Post-Uruguay Round Tariff Regimes – Achievements and Outlook.* Paris.
OJEC (1992) L180 and L181, 1 July, Brussels.
Page, S. and Hewitt, A. (2002) 'The new European trade preferences: does 'Everything But Arms' (EBA) help the poor?', *Development Policy Review,* 20: 1, 91–102.
Roningen, V.O. and Dixit, P. (1989) *Economic Implications of Agricultural Policy Reform in Industrial Market Economies* (U.S. Department of Agriculture, Economic Research Service, Washington DC).

Senti, R. and Conlan, P. (1998) *WTO – Regulation of World Trade after the Uruguay Round* (Zürich: Schulthess Polygraphischer Verlag).

Sheehy, S.J. (1995/96) 'Options for CAP Reform in the next trade round – with special reference to milk', *Proceedings of the Agricultural Economics Society of Ireland,* 111–34.

Sheldon, I.M. (2002) 'Regulation of biotechnology: will we ever "freely" trade GMOs?', *European Review of Agricultural Economics,* 1st quarter, 29: 1, 155–76.

Swinbank, A. (1999) 'CAP reform and the WTO: compatibility and developments', *European Review of Agricultural Economics,* 26: 3, 289–407.

USTR (1990) *Office of the United States Trade Representative.* October, Washington.

Valdis, A. and Zietz, J. (1980) *Agricultural Protection in OECD Countries: Its cost to Less-Developed Countries,* International Food Policy Research Institute, Research Report 21, Washington DC: IFPRI, December.

Viner, J. (1950) *The Customs Union Issue* (New York: Carnegie Endowment for International Peace).

World Bank and OECD (1993) *Trade Liberalization: Global Economic Implications.* 26 September (Washington and Paris).

Yeats, A. (1981) 'The influence of trade and commercial barriers on the industrial processing of natural resources', *World Development Report,* 9, 485–94 (Washington: World Bank).

Conclusions: The Future of EU and World Agriculture – The Challenges Ahead

Chapters 1, 4, 5, 7 and 8 have shown the extent to which the EU is a major global actor, not only in terms of food production and trade, but also with regard to its presence on the scene of modern research, that is of biotechnology. For a long time, the EU agricultural system has been portrayed essentially as a closed system favoring the interests of its own farmers, particularly the wealthiest therein, at the expense of the other agricultural systems in the world, particularly those of developing economies. In the last decade, the EU has multiplied the number of trade agreements with the developing world, strengthening its commitment to economic development in poor nations. The EU view that developing countries should be accorded special attention and that they should benefit from a special treatment going beyond Lomé, has been vehemently expressed in the recently opened WTO negotiations at the Doha Conference. This laudable standpoint differs from that of the other main agricultural producer and trader in the world, that is, the USA. Moreover, and partly because of outside pressures, the EU has made its CAP more compatible with the world agricultural trading system over the last ten years. Economic rents in the highly protected EU agricultural system have tended to weaken: export refunds have started being dismantled and producer prices have started to converge towards world prices making the CAP of today more market-oriented than it has ever been before. Even though the increasingly larger compensation payments paid to EU farmers have only changed the nature of support, as opposed to its philosophy, the CAP represents a declining share of the EU budgetary expenditure.

The essential question facing modern agriculture, from a European perspective, is indeed the following one: how can the EU wholeheartedly contribute to global food security and to the alleviation of poverty and malnutrition in the world, while preserving the environment? In other words, how can the CAP of the 21st century be compatible with – and geared towards – the crucial aim of raising standards of living in the developing world, without compromising the well being of generations to come?

Besides its numerous trade agreements with developing countries, an important direction taken by the EU in recent years has been to enhance its research capability and to collaborate with these countries so as to allow technological transfer to take place (see for example CEC, 1997). However, we have demonstrated that research alone is not sufficient to guarantee that the noble objective of eliminating starvation in the world be achieved. Also, the experience of the 'new style' agreements signed recently with ACP countries (the Cotonou Convention) and with the Mediterranean world shows that here again the EU agricultural 'sensitive' products tend to be excluded from these agreements. Yet, and as was highlighted in Chapter 1, the changing comparative advantages of a country reflect its welfare improvement. Also, as we have discussed in Chapter 7, many developing countries, including some from Africa, have a comparative advantage shifting towards meat and dairy commodities, two sensitive sectors for the EU. Consequently and illogically, the EU 'new style' development policy still inhibits these countries' possible trade gains, and consequently, their further economic development. The situation for the poorest developing countries is even more critical. The benefits derived by the developing world from global trade liberalization (that is from the UR) have been extremely uneven, with the share of the poorest countries in world trade diminishing substantially. It seems therefore that the EU ambition to play a greater role in reducing poverty and malnutrition throughout the world is still fraught with many contradictions.

It is clear that an unlimited and unconditional involvement of the EU in the task of enhancing both food security worldwide and the preservation of the environment is becoming urgent, given the challenges facing world agriculture in the years to come.

The challenges in the future

The challenges lying ahead are of two broad kinds. The first type refers to the challenges specific to the EU itself, such as the enlargement to Eastern Europe, and the geographical polarization of EU agriculture. To these could be added a number of long-standing issues, such as the problem of farmers' incomes (the 'farm problem'), and the continuing high prices faced by EU consumers. The second one is of a more global nature, as it encompasses world hunger and food security, water supply, the unknown long-term effects of biotechnology progress and food quality.

With regard to the EU proper, Chapter 8 has shown that, in the case of consumer prices, the optimistic expectations and forecasts of the late 1980s and early 1990s on the impact of agricultural trade liberalization did not materialize. Consumer prices have kept increasing in the EU since 1995, with the exception of Austria, Finland and Sweden, three countries that joined the EU in 1995. This increasing trend went in parallel with continuing decreasing producer prices over the period, giving the intermediaries, in particular the distribution sector, the ability to boost their profit margins. The problem of farmers' incomes has been addressed in the 1992 CAP reform, through proposals for diversification and early retirement schemes (Chapter 4). This problem will however be exacerbated in the near future, given that a number of relatively poor countries, with large and inefficient agricultural sectors, will be joining the EU. Chapter 6 has dealt with the issue of enlargement, which is dominated by the principle of budgetary discipline. It is imperative that the problem of declining numbers employed in the agricultural sector, coupled with increasing consumer prices be given special attention by the governments of applicant countries.

In the same way as imperfect competition in the distribution sector prevents the transmission of gains arising from agricultural trade liberalization to final consumers, EU agriculture tends to depart increasingly from the ideal competitive framework once formalized by Léon Walras. In particular, EU agricultural production is gradually more and more polarized from a geographical standpoint. According to the Bruges Group (1996), 70 to 80 per cent of European agricultural production in volume terms could become regionally polarized along the coast, stretching from Western Brittany to North-East Denmark, by 2010. As seen in Chapter 6, food research, in particular biotechnology research is already extremely polarized regionally, and it is enormously concentrated from an industrial economics perspective. Given that biotechnology research is now a fundamental component of competitive advantage in the agricultural and food sectors, the problem of peripheralization of both the new applicant economies in Europe and of poorer nations in the world is growing.

At a more global level, the peripheralization of food research mitigates against the objective of solving once for all the problems of world hunger, poverty and food security. These are by far the most important issues. In Chapter 5, we have warned against the danger that today's biotechnology research and development, in contrast with green revolution technologies in the past, is mostly carried out by private business organizations. These private firms are motivated more by the short-term objective of profit maximization rather than by the ideal of a society's welfare optimization. The impact of GMOs on solving the problem of hunger and malnutrition in poor countries is likely to be extremely marginal in the years to come. As a result, the significance of world hunger, poverty and food security may intensify in the near future, and this is also explained by the following two trends: increasing population and diversification of food consumption in fast growing economies. First, and according to FAO forecasts, population increases from the current 6 billion to more than 8 billion in 2025 require a doubling of food production in the next 20 years or so. Even if biotechnology advances were to offer the providential expected solution to population increases, we have highlighted that the major problem so far has been one of less sufficient food production at the global level, than one of equitable distribution of food resources (Chapter 1). The rich countries of the world, in particular the EU and the USA, produce enough food to feed a growing world population, and yet malnutrition and famines still afflict 850 million people in the world today. More disturbingly, malnutrition has made a reappearance in the countries that managed to become the bread baskets of the world, such as Argentina. Second, although developing Asia has been increasing its share of world food production, the Asian consumer shift from basic food commodities, such as rice, to more luxurious products such as eggs, poultry, red meat and beer, places an additional strain on world agriculture. This shift requires that ever growing quantities of cereals and sweet potatoes be produced for the use as animal feed. As a result of this exponential increase in the demand for cereals, it is estimated that a country such as China is gradually shifting from being the major world producer of cereals to becoming the biggest world importer of the same. Forecasts predict that by 2030, China will have to import an amount of cereals in volume terms, which is in excess of total world exports of the mid-1990s level.

Boosting world food production needs to be done while minimizing all economic externalities (such as soil erosion), and by overcoming water and resource shortages.

With a world population that has tripled during the last 70 years, water use has grown six-fold. Today, more than 500 million people live in countries characterized by water stress or scarcity, and it is forecast that by 2025, this number will reach 3 billion people (PANOS Institute, 2002). Again, this problem affects mostly the southern hemisphere of our planet, reinforcing therefore the risk of agricultural production polarization. As emphasized by the Asian Development Bank, future agricultural production is jeopardized by the availability of water worldwide. Pollution, forest degradation, and increased agricultural, domestic, and industrial use contribute to making both the quality and quantity of water decline (ADB, 2001). China illustrates again the case of a fast developing and large economy where the poor management of natural resources represents a limiting factor in agricultural production. The proportion of cultivated area in rural China irrigated with untreated water declines as the use of pesticides increases; farms and homes are often contiguous to industrial plants, allowing soil contaminants to move quickly along the food chain (Chemical Engineering, 1995; World Bank, 1995). A great deal of these challenges are being addressed today by research in the area of biotechnology, through the development of new crop varieties with high yield potential and high water use efficiency. However, this poses another challenge to future European and world agriculture: how safe are biotechnology advances?

The unknown long-term effects of genetically modified organisms on human health and on the environment are indeed one of the greatest challenges in the future. Understandably, no study to date has ever been conducted to show that the long-term negative effects of GMOs – such as the global 'contamination' of all crops by genetically engineered organisms (see Chapter 5) – will be greatly compensated for by the solutions that biotechnology research brings to the problems of both poverty reduction, and of the natural limits on food production imposed by pollution and soil erosion. It is unfortunately not possible for any scientist at the moment to demonstrate that genetically engineered organisms can be considered as one of the most wonderful innovations of humankind, and that they pose only minor risks to humans' health, risks that are greatly compensated by enormous gains for society as a whole. As we have seen in Chapter 5, these issues have taken a particular resonance in the context of the EU.

Finally, implementing the CAP in 1962 was an extremely successful initiative in that the policy did reach an important and chief objective stated in article 39 of the Rome Treaty, which was to provide regular supplies of food products to EU consumers. The same objective ought to be attained at the global level in the future, obviously with different instruments. The commitment of the EU, reiterated at the Doha conference, to promote the integration of the least developed countries in the world economic and trading system, should be a first step towards the design of a new *humanist agricultural and economic architecture.* The least developed economies of the world have had virtually no voice so far in WTO deliberations. The specificity of agricultural markets (described in Chapters 2 and 3), a manifestation of which is the polarization of agricultural activity as stressed above, as well as uneven levels of development throughout the world, invite extreme caution with regard to further multilateral agricultural trade liberalization. For most EU consumers, as for many other consumers around the world, food safety and food quality are of paramount importance.

Further reading

ADB (2001) *Water for All: The Water Policy of the Asian Development Bank* (Manilla: Asia Development Bank).

CEC (1997) *The European Initiative for Agricultural Research for Development.* Communication to the Council and to the European Parliament, COM (97) 126 final, Brussels.

Chemical Engineering (1995) *China Sets a New Agenda,* April, 30–35.

Groupe de Bruges (1996) *Agriculture, un tournant nécessaire* (La Tour d'Aigues: Editions de l'Aube).

PANOS Institute (2002) *Food for all.* Annual Report, London.

World Bank (1995) *China – Macroeconomic Stability in a Decentralized Economy* (Washington: The World Bank).

Index